THE RISE OF MAMMALS IN AFRICA

THE RISE OF MAMMALS IN AFRICA

Emmanuel Gheerbrant

ILLUSTRATED BY
CHARLÈNE LETENNEUR
AND ALEXANDRE LETHIERS

JOHNS HOPKINS UNIVERSITY PRESS

BALTIMORE

© 2025 Johns Hopkins University Press
All rights reserved. Published 2025
Printed in the United States of America on acid-free paper
9 8 7 6 5 4 3 2 1

Johns Hopkins University Press
2715 North Charles Street
Baltimore, Maryland 21218
www.press.jhu.edu

Library of Congress Cataloging-in-Publication Data is available.

A catalog record for this book is available from the British Library.

ISBN 978-1-4214-5239-5 (hardcover)
ISBN 978-1-4214-5240-1 (ebook)

Many of the illustrations in this book were drawn by Charlène Letenneur or Alexandre Lethiers.
They are credited in the captions of those figures that feature their work.

*Special discounts are available for bulk purchases of this book. For more
information, please contact Special Sales at specialsales@jh.edu.*

EU GPSR Authorized Representative
LOGOS EUROPE, 9 rue Nicolas Poussin, 17000, La Rochelle, France
Email: Contact@logoseurope.eu

La paléontologie doit tendre à retracer les phases
de peuplements des différentes régions du globe
et à expliquer la distribution géographique des êtres vivants.

(Paleontology must aim to retrace the phases
of faunal settlements in different regions of the globe
and explain the geographic distribution of living organisms.)
—CAMILLE ARAMBOURG, 1936

Contents

Color plates appear following page 42

Summary

The evolutionary explosion of mammals at the beginning of the Cenozoic Era, which began 66 million years ago, is a key event in our distant origins, and it is one of the main topics in current biological and paleontological studies. Knowledge about the origin of the large living groups of marsupials and placentals has made considerable progress following recent important discoveries in two different disciplines of the life sciences, paleontology and molecular biology. The latter has confirmed the role of Africa as the cradle of several large modern groups of placental mammals such as the afrotherians, which include a third of the extant mammalian orders, among which are elephants. Important discoveries of the former also demonstrate the role of Africa in the early evolution of modern mammals, especially for African ungulates (elephants, hyraxes, sea cows) and primates (lemurs, monkeys, apes). Paradoxically, this African evolution took place in a singular paleogeographic context, separated from other continents.

This was "Island Africa," a gigantic island continent that remained isolated south of the ancient Tethys Sea for more than 80 million years, from 110 to 23 million years ago. Island Africa was not completely isolated, however. In fact, its isolation was repeatedly broken. This is biologically illustrated by a succession of episodes of mammal dispersals from Eurasia across the Tethys Sea. Various Eurasian mammals arrived at different times throughout the early Tertiary in Island Africa, where they established themselves. They had great evolutionary success in Africa, giving rise to several remarkable endemic radiations. Island Africa was the cradle of an important number of new groups of mammals, including several extant groups present in modern fauna. Some of them, elephants and monkeys, for instance, are famous and beloved; others, such as hyraxes, aardvarks, elephant shrews, otter shrews, springhares, and flying squirrels, are little known to most people, but they constitute an important and remarkable part of the endemic fauna of Africa today.

This book outlines the fossil history of ancient African mammals and their evolution in light of the most recent discoveries and through a review of paleontological knowledge and field research.

Preface

What is the origin and ancient evolutionary history of modern mammals in Africa, beginning from Africa's oldest paleontological roots? What fossils do we have? What is the importance of the African cradle in the evolution of the world's fauna and present continental biodiversity? What is the geographic and environmental context of ancient African evolution in a world quite different, at that time, from ours?

These questions about the ancient evolution of African mammals are of direct concern to us—as humans and members of the ape family—but they also concern many other extinct and present-day animals. Some, like elephants, are emblematic today, although what is little known is that elephants' existence is almost entirely in the past, since only three living species remain of the 180 fossil species recorded since the beginning of the "Age of Mammals." Other mammals of the ancient African cradle, such as arsinoitheres, dassies, elephant shrews (sengis), springhares, and African "flying squirrels," are more mysterious, but they too flourished. The rich and long evolutionary history of African mammals is largely unknown outside a small community of specialists. This is so despite many recent paleontological discoveries and 150 years of field research as well as key discoveries of molecular biology.

In exploring these discoveries and tracing this history, this book tells, for the first time, the story of the ancient origins of modern mammals in Africa. It shines light on the particular conditions, and extended period of 80 million years, when these mammals evolved on a gigantic island continent. The roll call of placental mammals that arose in the African cradle is a long one: monkeys, galagos, elephants, dassies, elephant shrews, aardvarks, hyaenodont carnivores, flying squirrels, springhares, mole-rats, and many others. From the earliest-known fossils to the birth of the Old World and the establishment of the modern fauna, let us embark on a journey through the long history of a huge lost world, Island Africa!

Acknowledgments

This book follows more than 30 years of research on the early evolutionary history of mammals in Africa, since my PhD thesis in 1989, which was devoted to the study of one of the oldest-known placental mammal faunas in Africa (at the Adrar Mgorn sites, Morocco). It is primarily based on a database that inventories the mammal species reported in the African Paleogene fossil localities, which I began to establish for my HDR (Habilitation à Diriger des Recherches) thesis (1999) and which I have continued to enrich and update with new discoveries over the ensuing years (see appendix). The conception and birth of this book were the outcome of stimulating collaboration and discussions with many colleagues from the French and international paleontological community.

Among these colleagues, Jean-Claude Rage (Centre National de la Recherche Scientifique, or CNRS) deceased, sadly, in 2018; I remember our deep friendship and our many friendly discussions that enormously enriched my understanding of the paleobiogeographic history and evolution of African and Gondwanan faunas. I must pay him my greatest tribute here. As my PhD thesis advisers, Jean-Jacques Jaeger and the late Bernard Sigé guided my initial steps in the fields of paleontology and evolutionary history of African mammals. At the beginning of my career, Eric Buffetaut (CNRS) stimulated my studies in vertebrate paleontology at the Paris 6 University (now Sorbonne University), especially on questions of vertebrate paleobiogeography. Also at the Paris 6 University, the late Luc-Emmanuel Ricou and Jean Dercourt introduced me to the knowledge of the paleogeographic and geodynamic context, particularly in Tethyan and peri-Tethyan regions, which has been instrumental in my understanding of its significance for the paleobiogeographic history of continental faunas (see Gheerbrant, 1987). Philippe Janvier (CNRS) spared no effort to enable the paleontology team from the University of Paris 6 (today Sorbonne University), of which I was a member, to be attached to the Muséum National d'Histoire Naturelle (MNHN) in 1999. Philippe has taken an interest in my work since this time, including after his retirement. I am grateful to him for reading of the first version of the manuscript for this book and for editing the English.

Pascal Tassy and Ken Rose helped me to advance the book project, including helping with the search for a publisher. Thanks to Ken especially for directing me to the editors of Johns Hopkins University Press. I am grateful to Martin Pickford (formerly Collège de France, but now "Attaché Honoraire" at the MNHN), who kindly accepted the thankless and important task of reading and editing my poor English. I also thank him for sharing his vast knowledge of African mammals and, in particular, for information concerning the exceptional Namibian mammal localities that he discovered.

I thank Guntupalli V. R. Prasad (University of Delhi) for discussions on the systematics, evolutionary history, and paleobiogeography of mammals in Gondwanan continents. I owe to Herbert Thomas (Collège de France) the extension of my paleomammalian studies to the Arabian faunas from the famous Omani sites of the Dhofar region that he discovered with Sevket Sen (CNRS) and Jack Roger (BRGM). Thanks to him for inviting me to participate in the field paleontological excavations in the Taqah locality and in the beautiful Omani desert of Thaytiniti. My study of the Omani mammals began during a postdoctoral research stay kindly hosted and supervised by Philip D. Gingerich in Ann Arbor at the Museum of Paleontology, University of Michigan. This postdoctoral position allowed me to learn a lot about the great evolutionary history of African primates, thanks especially to the fantastic collection of casts assembled by Philip. William (Bill) J. Sanders and our late colleague and friend Greg Gunnell (both of the Museum of Paleontology, University of Michigan) gave me a lot of encouragement to study African mammals, and Bill was one of the first colleagues to encourage me to write a book on the early evolutionary history of mammals in Africa.

What is known today about the paleontology and evolutionary history of African mammals is the result of field research carried out by many renowned paleontologists since the end of the 19th century, a brief account of which can be found in this book. I had the immense luck to be able to follow the great scientific and human adventure of the paleontological exploration of Africa, which led me to new discoveries reported and illustrated in this book. My paleontological field research was made possible thanks to the critical help and participation of many colleagues, too numerous to mention all by name, during the surveys and excavations, in particular in Morocco. I must pay tribute to them here. My field surveys were financed by several institutions, in particular the Centre National de la Recherche Scientifique (Direction des Relations Internationales, Cooperation Agreements CNCPRST and PICS programs), the Muséum National d'Histoire Naturelle (BQR and ATM funding programs), and the National Geographic Society.

I thank Laurent Marivaux (CNRS, Université Montpellier II) for his valuable insights into the systematics and evolution of African rodents. I thank Rodolphe Tabuce (CNRS, Université Montpellier II) for enlightening discussions and information about the evolution of early African hyracoids and macroscelideans, and for providing many working documents such as casts and photographs. I thank Floreal Solé for valuable insights into the systematics and evolution of African hyaenodonts. I would also like to thank Robert J. Asher (University of Cambridge) and Erik R. Seiffert (University of Southern California), who provided me with a lot of information on Paleogene mammals from Africa (particularly from the Fayum) and who kindly answered my questions about the groups they study.

Many colleagues have graciously agreed to provide beautiful photos and illustrations, some of them unpublished, which have enriched and embellished this book: K. C. Beard (plate 7 lower), P. D. Gingerich (plate 13 lower), P. Loubry (plate 14), L. Marivaux (fig. 4.15, plate 2 upper, plate 13 upper), M. Mathison (plate 1 lower), M. Pickford (plate 7 upper, plate 12), S. Sen (plate 15A), F. Solé (fig. 4.13A), R. Tabuce (figs. 4.8A, 4.13A, and 4.14; plates 3, 6, and 11), and I. Zalmout (plate 16). I thank P. Loubry for the photos of the skeletons of *Saghatherium antiquum* (plate 14) and for his help in processing and formatting the photos (plate 15 and others). I thank L. Cazes for the photos of the skull of *Aegyptopithecus* (fig. 4.30). Some pictures were taken in exhibitions and museums: MNHN–Galerie de Paléontologie (Paris) and the Geological Museum of Cairo (skeleton of *Moeritherium* photographed by M. Pickford and redrawn in fig. 4.17A by A. Lethiers). Charlène Letenneur (drawings of fossils and restoration of animals) and Alexandre Lethiers (graphics, tables, maps, drawings) have combined their talents to give this book attractive and informative illustrations. The book's illustrations by Charlène and Alexandre are indeed the best invitation to anyone who wants to discover the past diversity and evolution of mammals on the African island. I would like to thank the entire MNHN Paleontology Laboratory (current abbreviation: CR2P), where I have been able to do my paleontological research over the past 25 last years in a warm and fruitful milieu, and in the wonderful environment of the Paris Museum, with its collections, galleries, garden, and historical coverage exceptional for natural sciences, and with its great human community. I would also like to express my gratitude to the entire French university community, which provided me with my scientific training and paved the way for me to become a paleontologist.

I dedicate this book to my beloved wife, Bénédicte, and to my very dear two children, Ella and Nathan, without whom this book and many other things would not have existed. They have provided much of my life's energy and allowed me to realize my passion for paleontological research. Special thanks also to my dear mother-in-law, Christiane, who never stopped encouraging me in the writing of this book.

I am grateful to the editors at Johns Hopkins University Press, in particular Tiffany Gasbarrini, for taking an interest in my manuscript and for supporting its publication. I would also like to thank Ezra Rodriguez, Robert Brown, and Carrie Love for help with editing and formatting the manuscript and the figures and with the final assembly of this book.

THE RISE OF MAMMALS IN AFRICA

The Living Fauna of African Mammals

The African "wild world," in which humans took their first steps, has fascinated and fed our collective imagination since antiquity. Today, it is one of the greatest natural heritage sites of the world. Mammalian megafauna is the most emblematic of African wildlife, both for its richness and for its spectacular species. It is more abundant and diversified than that of other continents that were decimated by the Quaternary extinctions, mostly between 130,000 and 13,000 years ago. The African megafauna is famous, especially the iconic "Big Five": the elephant, lion, rhinoceros, leopard, and buffalo **(fig. 0.1)**. As the price of their iconic status, they have been preferred targets during African safaris. But, fortunately, they are also flag bearers for the great natural reserves and national parks of Africa listed as UNESCO World Heritage Sites, such as those in Niger, Kenya, Botswana, the Congo, and South Africa. At present, the Big Five account for only a small fraction of large African mammals, which include many other popular species such as giraffes, zebras, hyenas, chimpanzees, gorillas, monkeys, hippopotamuses, antelopes, and warthogs.

In Africa, one also finds a great number of mammals of smaller dimensions. These African mammals are not as well known but are just as intriguing and remarkable. They include galagos, okapis, flying squirrels, naked mole-rats, springhares, pangolins, aardvarks, dassies, African otters (otter shrews), elephant shrews (sengis), golden moles, and fruit bats **(fig. 0.2)**.

In fact, the mammalian fauna of Africa is the richest in the world in number and in variety. It counts 1,150 species in 57 families and 13 orders,

Figure 0.1. The Big Five of living African mammals (*left to right, from top*): elephant, lion, rhinoceros, leopard, and buffalo. Of these, only the elephant has ancient evolutionary roots on Island Africa. Drawing by C. Letenneur.

all of them placentals (the largest group of extant mammals and the one to which we belong). That is about 20% of all extant mammal species. As such, Africa represents a major hot spot of continental biodiversity. There are several geographic reasons for this. Africa is the second-largest continent in the world, and it is the largest in the tropics—and tropical habitats are a source of high biodiversity. Africa is thus both a major continental center of evolution and a large contemporaneous reservoir of biodiversity.

Figure 0.2. Diversity of living African mammals (*left to right, from top*): African otter shrew (*Potamogale*), elephant shrew (*Rhynchocyon*), African civet (*Civettictis civetta*), aardvark (*Orycteropus*), springhare (*Pedetes*), African flying squirrel (*Anomalurus*), dassie (*Procavia*), cane rat (*Thryonomys*), naked mole-rat (*Heterocephalus*), gorilla (*Gorilla*), and bushbaby (*Galago*). Drawing by C. Letenneur.

Some groups, orders and families, have exceptional diversity on the African continent. Anthropoids, ungulates, carnivores, and rodents have among the greatest number of species, encompassing diverse shapes, sizes, and adaptations. For example, there is an incredible variety of monkeys, "antelopes" (gazelles, dik-dik, oryx, duikers, wildebeest), foxes, mongooses, genets, rats, porcupines, squirrels, and bats. The diversity of small mammals is remarkable in Africa, especially among rodents, bats, and shrews. The diminutive shrews of the genus *Crocidura* alone account for more than 100 species in Africa!

The modern biodiversity of Africa is the result of a long history, not only of local, or endemic, evolution but also of ancient intercontinental faunal interchanges linked to the tectonic history of the Arabo-African plate since the time of the ancient supercontinent Gondwana. This means that most extant African mammals, including four of the Big Five, are not native to the continent. They are "recent" arrivals, on a geological timescale, that is. These immigrants arrived after the collision of Africa and Eurasia 23 million years ago (or Ma, for mega-annum). Only one of the Big Five, the elephant, has a long autochthonous, or indigenous, history in Africa that goes back to the beginning of placental evolution, more than 60 Ma—that is, 60 million years ago.

Six of the 13 orders of placentals known today in Africa are indeed recent colonizers of the continent: carnivores (lions, cheetahs, hyenas, jackals, mongooses), perissodactyls (even-toed ungulates such as zebras and rhinoceroses), artiodactyls (odd-toed ungulates such as antelopes, giraffes, and buffalo), rabbits, rodents such as rats, and insectivores such as shrews and hedgehogs. Some of them, such as artiodactyls and carnivores, are the most diversified mammals of extant African faunas. Their ancestors arrived in a great southward dispersal from Eurasia when the isolated Arabo-African plate collided with Eurasia some 23 Ma. They quickly adapted and spread through the favorable tropical environments of Africa. They flourished in diversity to the point of forming a large part of the richness of African fauna today. Indeed, evolution in Africa played a major role in the rise of modern mammals and their current diversity.

This evolution also concerns us more directly, of course. In the long African evolutionary history of mammals, one of the last events is the rise of humans and their forebears (*Australopithecus*, 4–3 Ma; *Homo*, 2 Ma; *H. sapiens*, 150,000 years ago), whose "out of Africa" conquest of the world began 3–2 Ma, long after the beginnings of primates and anthropoids on the continent, in the Paleocene (57 Ma) and Eocene (41 Ma), respectively.

CHAPTER 1

Origin and First Diversification of Placental Mammals in the World and in Africa

The Cenozoic (or Tertiary), from 66 Ma to the present, is the "Age of Mammals." It began after the disappearance of the non-avian dinosaurs, with the evolutionary explosion of the modern mammals, during which the 18 extant orders of placentals—including primates, bats, rodents, rabbits, horses, cows, elephants, and carnivorans—originated. The placentals, to which we belong, are one of the three major living groups of mammals, the others being the marsupials (pouched mammals, such as the kangaroo) and monotremes (egg-laying mammals, such as the platypus). Of these three groups, the placentals are by far the most diverse. The key event in the evolutionary explosion (also known as an evolutionary radiation) of modern mammals following the extinction of the dinosaurs is at the heart of research in several scientific disciplines.

The initial explosive diversification of modern mammals

The explosion of mammalian diversity raises several big questions: What were the early relatives of the extant mammals at the outset of their evolution? What was the age of the first divergence of the major modern lineages? And what were the environmental (paleogeographic, climatic) and biological (extinctions, ecology) drivers during their early evolution? Our knowledge has progressed enormously toward answering these questions, following important recent discoveries in two major disciplines: (1) paleontology, which studies fossils by comparative anatomy as direct milestones of past evolution, and (2) evolutionary molecular biology, which traces the phylogeny of present-day species through the analysis of the cellular genome (DNA, RNA, and protein sequences). During the past 30 years, molecular phylogeny—a recent branch of the biological sciences—has radically changed our understanding of the higher relationships within mammals, particularly for the extant placental orders. It has demonstrated the existence of three major lineages, or clades, comprising several placental orders. Three superorders split early in placental history, and they originated and evolved in distinct continental provinces **(fig. 1.1)**: (1) the xenarthrans, including the armadillos and sloths, of South America, (2) the afrotherians of Africa, which comprise the oldest African placental lineages, such as elephants and elephant shrews, and (3) the boreoeutherians, including euarchontoglires and laurasiatherians, which gather all placentals originating from Laurasia (northern continents: North America and Eurasia), such as horses, cows, primates, rodents, and carnivores.

The content of these superorders—in other words, which orders are related to them—is now fairly well understood and defined, but their relationships to each other remain poorly resolved and much debated. For example, some molecular phylogenies place afrotherians (elephants, sengis) and xenarthrans (armadillos) in a single group—a "super-super" order—called Atlantogenata (meaning literally "born on both sides of the Atlantic"); they consider atlantogenatans and boreoeutherians the first lineages to split in the evolution of placentals (Barrier et al., 2008). However, there are two other competing hypotheses in which the first branch of placentals to split is either the Afrotheria or the Xenarthra. It is notable that all these hypotheses imply an ancient origin of the Afrotheria during the evolution of placentals.

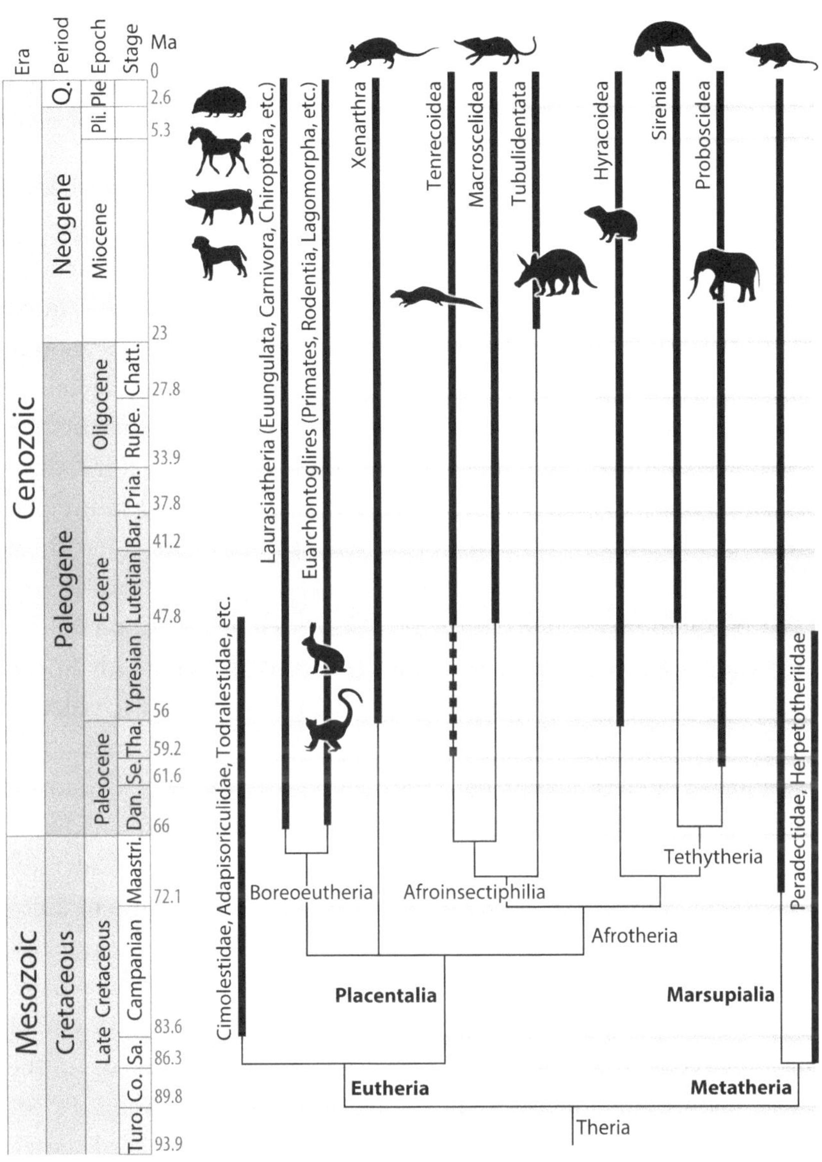

Figure 1.1. Relationships and stratigraphic distribution of the higher placental groups with details of the afrotherian orders. The *thicker lines* indicate the stratigraphic extension of the clades based on known fossils; the *thinner lines* indicate the relationships between the clades and their possible temporal extension inferred from their relationships. Graph by A. Lethiers.

When, why, and where did the mammalian explosion occur? Contradictions and reconciliation of molecular and fossil evidence

The dating of the origin and first diversification of mammals has been a subject of lively debate between paleontological and molecular researchers. The dispute is mainly about the role of the mass extinction crisis at the Cretaceous–Paleogene boundary (66 Ma)—the famous "KPg crisis" (formerly known as "KT crisis")—in the diversification of modern mammals. The role of the KPg crisis in the early placental explosion is decisive for paleontologists, but not for all "molecularists."

Those who study fossils almost unanimously conclude that there was an evolutionary explosion of modern mammals with rapid splitting of the main placental orders at the beginning of the Cenozoic, mostly in the Paleocene, between 66 and 56 Ma, with an evolutionary peak in the first five million years. This major event in vertebrate evolution follows the mass extinctions at the end of the Cretaceous, during which 76% of the Cretaceous species disappeared (Barnosky et al., 2011), in particular all the non-avian dinosaurs. For paleontologists, the evolutionary diversification of modern mammals at the beginning of the Cenozoic corresponds to an "adaptive radiation." This was an opportunistic and ecological phenomenon related to the conquest by mammals of adaptive niches (e.g., nutritive niches) left vacant by the disappearance of dinosaurs, in particular the herbivorous niches that were not free for mammals during the dinosaur era. In other words, the rise of modern mammals is a "recovery event" from the great extinction crisis at the end of the Cretaceous (the KPg crisis). This is consistent with the age of the oldest-known fossils of modern mammals (dating to the earliest Paleocene) and with the evolutionary state of the species that succeeded each other during this early period of diversification.

Several teams of paleontologists have defended the radical view of an abrupt adaptive radiation of placentals, starting from the early Cenozoic (66 Ma). This is a "hard" explosive model (one of the two variants of the explosive model) of the evolutionary radiation of modern mammals. For the North American team of J. R. Wible (2007), the detailed study of the

Cretaceous mammal relationships indicates indeed that there are no known fossils of early placentals older than the Cenozoic. For Wible and colleagues and more recent researchers (Goswami et al., 2011; Halliday et al., 2017; O'Leary et al., 2013), Cretaceous mammals, including the most derived ones, are in fact all ancestral groups (stem eutherians) more or less closely related to the placental root but not directly related to extant placental orders. Nevertheless, we know a few probable Cretaceous true placentals. *Protungulatum* (66 Ma) is, for instance, identified as a laurasiatherian euungulate for some of the same paleontologists who advocate the hard explosive model (O'Leary et al., 2013). As a Cretaceous placental, *Protungulatum* supports an alternative "soft" explosive model of placental radiation, advocated by the Australian biologist Matthew J. Philips (Phillips and Fruciano, 2018), in which placental origins go back to a little before the Cenozoic, during the last stages of the Cretaceous, between 83 and 77.6 Ma in the estimation of Álvarez-Carretero et al. (2022). This agrees with paleontological studies that admit the existence of major placental lineages (i.e., large supraordinal groups) during the Cretaceous—lineages that are at the origin of the explosive placental radiation during the early Cenozoic. Interestingly, recent paleontological discoveries also envision an explosive model of radiation for birds at the beginning of the Cenozoic (Ksepka et al., 2017).

Research in evolutionary molecular biology has changed not only the way we look at the relationships of major groups of mammals but also their age of origin—based on the so-called molecular clock. Compared to paleontology, early molecular-clock dating yielded much older ages— from long before the Cenozoic—for the origins of modern mammals: It extended the age of all placentals and most of their orders back to the Cretaceous (Kumar and Hedges, 1998). Early molecular-clock dating linked the basal diversification of modern mammals not to the extinction crisis at the end of the Cretaceous but to major paleogeographic changes that occurred during the Cretaceous (Madsen et al., 2001; Murphy et al., 2001; Springer, 2004; Springer et al., 2011), such as separation of the South American and African continental plates 110–100 Ma. Hence the geographically derived names of the placental superorders—Afrotheria (mammals of Africa), Boreoeutheria (boreal mammals), Laurasiatheria (mammals of the northern continents). Thus, the molecular study by Bininda-Emonds and colleagues (2007) considered the radiation of modern placentals to be unrelated to the KPg crisis, and it indicates that the separation of the extant placental orders (i.e., the interordinal diversification) took place entirely during the Cretaceous—and at a fairly early age, around 90 Ma. This hypothesis corresponds to Archibald and Deutschman's (2001) "short fuse" model of evolutionary radiation. This model is a real bone of contention for paleontologists: In this hypothesis, the ages of the known fossils would not be significant at all, since there are no known fossils of placentals of earlier age than the Cenozoic. More recent studies of molecular phylogeny agree more with paleontological data—that is, with the age of fossils—for the temporal calibration of the evolutionary radiation of modern mammals. "Geomolecular" dating (Phillips, 2016; Phillips and Fruciano, 2018) best takes fossils into account to calibrate molecular ages. It traces the origin of placentals and some major lineages, such as laurasiatherians, euarchontoglires, xenarthrans, and afrotherians, to the latest Cretaceous, circa 80–75 Ma. It supports the soft explosive model of placental radiation mentioned above. In this model, both molecular and paleontological data are consistent with a scenario in which the early diversification of modern mammals is linked to the rapid conquest of the ecological niches left vacant after the KPg crisis, 66 Ma.

Fossils only indicate a most-recent date of appearance of lineages, but the lineages could go back further without a fossil record. The earliest-known fossils of placentals, with the exception of *Protungulatum*, date from the very beginning of the Cenozoic, some 66 to 63 Ma. They belong to several major lineages (orders), including insectivores (eulipotyphlans), carnivorans, primates, glires (the rodents and the rabbits), and ancestral euungulates ("condylarths" = stem groups of perissodactyls, such as horses, and of artiodactyls, such as pigs and ruminants). Even in Africa, where the fossil record is very incomplete, there is fossil evidence for an ancient divergence of living African ungulates, such as dassies and elephants, and for an origin of their common ancestor (i.e., basal paenungulates) before the Cenozoic. The early diversity of placentals and the relationships of their major branches indicate that their evolutionary history dates to the Cretaceous—this despite the absence of direct fossil evidence. Our fossil knowledge of the initial adaptive radiation of modern placentals is based on early Cenozoic species that are stem groups of superorders such as euarchontoglires, laurasiatherians, afrotherians, and paenungulates **(see fig. 1.1)**, and on a few others that are ancestral to extant orders such as primates, hedgehogs and shrews (eulipotyphlan "insectivores"), carnivorans, rodents, and elephants. The placental radiation has its ultimate origin in a Late Cretaceous, 80-million-year-old common ancestor. Knowledge of Cretaceous placentals remains one of the great challenges of field paleontological research worldwide, including in Africa.

The evolutionary radiation of modern mammals at the beginning of the Cenozoic was not only marked by a multiplication of evolving lineages but also by the geographic expansion of those lineages throughout the world. The first continental expansion of placentals, at the onset of the Cenozoic, took place in a paleogeographic context that basically involved dispersals (i.e., faunal migrations across continents). The age of the first placental lineages supports a mechanism of active dispersals of various stem groups across continents, rather than suggesting vicariance, which involves fragmentation of the geographic ranges of ancestral mammals by continental drifting (e.g., opening of the Atlantic Ocean between Africa and South America during the Cretaceous). Even in the hypothesis of a common ancestry of the Afrotheria and Xenarthra (i.e., the hypothesis of the clade Atlantogenata), the evolutionary divergence of these two extant groups implies dispersal between Africa and South America rather than vicariance, because these continents separated about 110–100 Ma—that is, *before* the origin of placentals. Ancestral placentals (stem groups) most likely dispersed from an Asian center of origin of therian mammals. After dispersal, they settled in new continental provinces, where they gave rise to major modern groups that evolved at the beginning of the Cenozoic: xenarthrans in South America, afrotherians in Africa, laurasiatherians and euarchontoglires in Laurasia (Eurasia and North America). Dispersals also explain the population of Australia by marsupials and of Madagascar by placentals in the Eocene. The colonization of the world by modern mammals thus largely results from active dispersals from a Eurasian center of origin, where the ancestral groups of placentals (eutherians) are the oldest. The modern mammalian explosion occurred with the multiplication of the geographic centers of diversification throughout the world.

What about Africa in this story?

The Cretaceous fossil record of the ancestral eutherian groups of placental mammals is geographically heterogeneous and contains many gaps. North America and Asia have yielded much more Cretaceous fossils of stem placentals than anywhere else in the world. Few discoveries have been made in the other continents. Some continents are even terra incognita, without any known Cretaceous fossils of stem (or crown) placentals. This is the case with the southern landmasses of Gondwana including Africa, South America, Madagascar, Antarctica, and Australia. The only known Cretaceous stem placentals from Gondwana were discovered from the latest Cretaceous (72–66 Ma) of India by the paleontologist G. V. R. Prasad of the University of Delhi (Prasad et al., 1994). There are a few poorly known species, represented by isolated teeth. Current knowledge suggests that paleontologists should look for early placentals in poorly known continental provinces. For instance, in 2010 the Californian paleontologist W. A. Clemens (1932–2020, University of Berkeley) showed that in North America, the continent with the best-known fossil record, the new species that appeared in the early Cenozoic were not the result of local evolution but were immigrants of unknown origin (Clemens, 2010).

What about Africa? Both molecular phylogenetics and the known fossil record indicate that Africa was likely an important center of origin and diversification of modern mammals. It was one of the major continental cradles of the initial placental radiation at the beginning of the Cenozoic—that is, one of the main places of the early evolution of the placentals. The African evolutionary radiation saw the rise and diversification of original groups that evolved with greater or lesser success, a number of them belonging to extant fauna. The African radiation is both geographic and ecological in nature. It corresponds to the colonization and population of Africa during the Cenozoic and to the occupation of ecological niches left vacant after the mass extinctions at the end of the Cretaceous, as in the other continents.

Identification by molecular phylogeny of the "supergroup" Afrotheria confirms that Africa was a major continental cradle in the evolution of placentals. Afrotheria includes one-third of the extant placental orders, nearly all being African in their early history. Their past diversity was actually far greater than that of the 90 known living species. There are two main clades (lineages) of afrotherians. One, called Afroinsectiphilia (meaning "African insectivores"), includes insectivorous-like mammals. Most of them are small afrotherians, and they include otter shrews and tenrecs, golden moles, elephant shrews or sengis (Macroscelidea), and aardvarks (Tubulidentata, the largest afroinsectiphilians). The other afrotherian group, called Paenungulata, consists of large ungulate-like mammals and includes sea cows (Sirenia), hyracoids (Hyracoidea, today represented by the small dassies), and elephants (Proboscidea). An important extinct lineage of paenungulates has also been discovered in Africa. This is the embrithopod order, best known by the amazing mammal *Arsinoitherium* from the Fayum Oligocene localities (34 Ma) in Egypt. *Arsinoitherium* was among the most remarkable mammals of the ancient African herbivorous megafauna. With the size of a rhinoceros, it was one of the prehistoric Big Five of the ancient African island.

Afrotherians have been found in Africa from as early as the beginning of the Cenozoic, in the middle–late Paleocene (between 60 and 56 Ma) and in the early Eocene (56–49 Ma), with fossils found of stem elephants, hyracoids, embrithopods, elephant shrews (sengis), and possibly of the tenrecoid insectivores (golden moles, otter shrews, tenrecs). Afrotherians were among the first evolving placental mammals on the continent. The origin of placentals in Africa remains enigmatic. This is because of large gaps in the fossil record. We do not know when afrotherians originated in Africa: When did afrotherians appear before their first-known fossils, dated at 60 Ma? Moreover, we do not know the composition of the African mammal paleocommunities in which the first afrotherians lived. What other groups and species did they live with during the Late Cretaceous and early Cenozoic? Did they coevolve at the same time with archaic, nontherian Mesozoic mammals and non-avian dinosaurs?

The early history of modern mammals in Africa is not restricted to that of afrotherians. Other extinct and extant lineages of placentals evolved early in the continent. First discoveries in the 1980s from the Paleocene of Africa recognized, for example, the oldest fossils of primates and carnivorous hyaenodonts in Morocco. These Paleocene primates and hyaenodonts are associated in the same faunas with enigmatic small insectivore-like species, identified in the late 1980s and early 1990s by the author (Gheerbrant, 1989, 1992, 1994, 1995), that document an early African evolution of stem and modern (crown) placentals. After the Paleocene, new groups successively appeared in Africa following several dispersal episodes from Laurasia during the Eocene and Oligocene, from 56 to 23 Ma; these groups included rodents (which arrived in two immigration waves), anthropoids, anthracothere artiodactyls, pangolins, and marsupials. After their arrival in Africa, these mammals evolved locally in an endemic way. The evolution in Africa of various lineages of different origins and the local extinctions (e.g., marsupials, adapiform primates, hyaenodonts, and embrithopods) have shaped the composition of extant African faunas.

CHAPTER 2

The African Theater of Origins

The paleogeographic history of the African continent was shaped by three major geodynamic events. The first two events were the opening of the Tethys Sea, around 170–150 Ma, and the opening of the Atlantic Ocean, around 110–100 Ma. This led to the separation and isolation of the Arabo-African plate from other continents during the Late Cretaceous and the Paleogene. The third event occurred at the end of the Paleogene, 23 Ma, with the intercontinental collision of Africa with Eurasia, which resulted in the establishment of the current Old World realm linking Africa, Eurasia, and India in a single, huge biogeographic continental province.

If we unroll the thread of geological time from the earliest periods, the ancient African faunas evolved in four distinct geographic worlds shaped by these geodynamic events (**fig. 2.1**): (1) Pangean Africa (earlier than Middle Jurassic, more than 170 Ma), (2) Gondwanan Africa (Middle Jurassic–Early Cretaceous, 170–110 Ma), (3) Island Africa (mid-Cretaceous–Paleogene, 110–23 Ma), and (4) the Old World (Neogene–today, 23–0 Ma).

A geography of origins apart: Island Africa, 110–23 Ma

Island Africa was one of the great continental theaters where modern mammals (including placentals) took their first steps. In fact, two-thirds

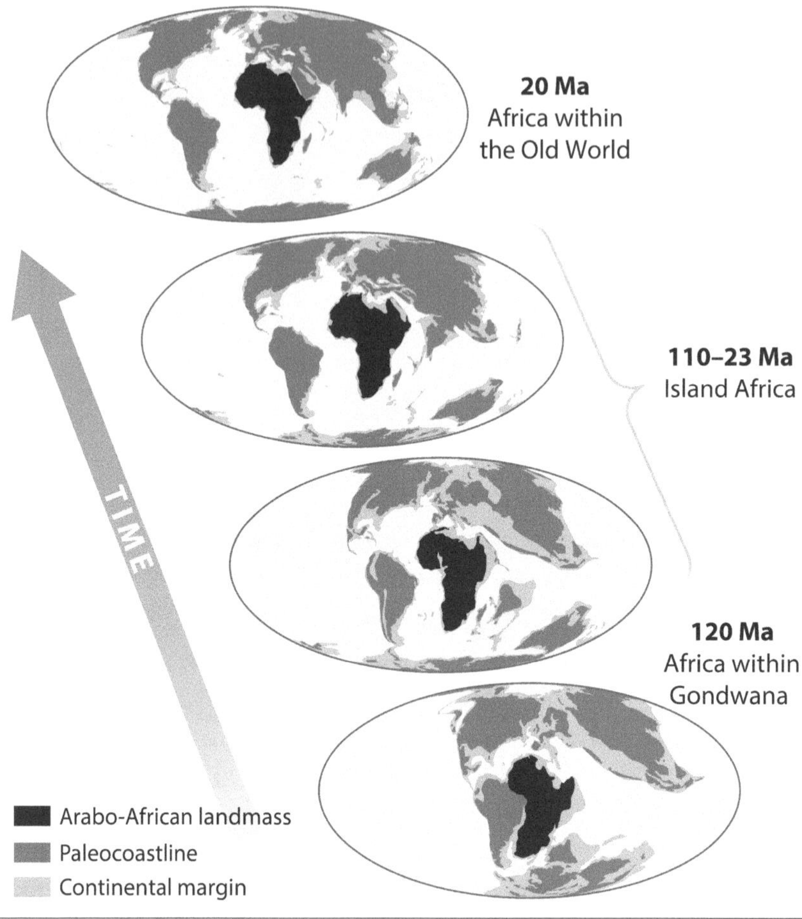

Figure 2.1. Paleogeographic evolution of Africa from Gondwana to the Old World. Redrawn by A. Lethiers from Kocsis and Scotese (2021, figs. 3–4).

of the history of modern African mammals took place in the geographic context of Island Africa. What do we know about Island Africa? It was a gigantic island continent that corresponded with the entire Arabo-African plate. It was one of the main landmasses resulting from the breakup of the Gondwana supercontinent; the others are South America, Antarctica, Australia, India, and Madagascar. It was isolated from South America from the mid-Cretaceous, 110–100 Ma (Albian), and from Eurasia from the Middle Jurassic, about 170 Ma, until the end of the Paleogene, 23 Ma (early Miocene). Island Africa existed for more than *80 million years.* During this long period, which was longer than the Cenozoic, the African fauna and flora evolved in an endemic way. This strongly shaped the African biota.

Island Africa formed a single vast continental area **(figs. 2.2 and 2.3)** that encompassed the Arabian Peninsula. This was due to the Red Sea not opening until the early Miocene, about 23 Ma. Some important paleogeographic changes nevertheless occurred in Africa. Field studies in the Saharan regions, especially those carried out in Nigeria by the Swedish ammonite paleontologist Richard Reyment (1926–2016), revealed the former presence of a vast ancient interior sea during the Late Cretaceous and the Paleocene that extended from north to south across western Africa (see Reyment, 1980; Reyment & Dingle, 1987). This was the so-called trans-Saharan sea. It separated an ancient western African domain from the rest of Africa. This African interior sea should have led to the existence of provincialism in the two distinct regions. However, provincialism in Island Africa remains poorly represented by fossil faunas

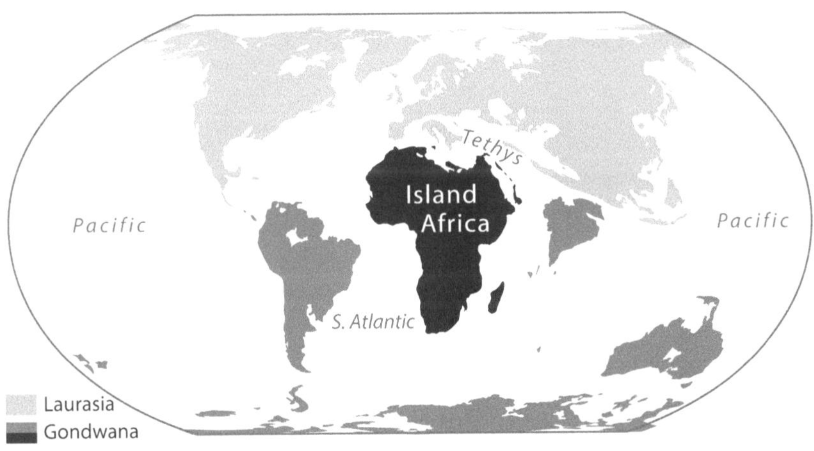

Figure 2.2. Paleogeographic map of the early Eocene world at 50 Ma. At that time and until the end of the Paleogene, about 23 Ma, Africa and the Arabian Peninsula formed a gigantic island continent: Island Africa. Note: For simplicity, the continental outlines do not include the continental shelf (e.g., the one that widely extends between North America and northern Europe). Redrawn by A. Lethiers from Scotese (2021, fig. 18).

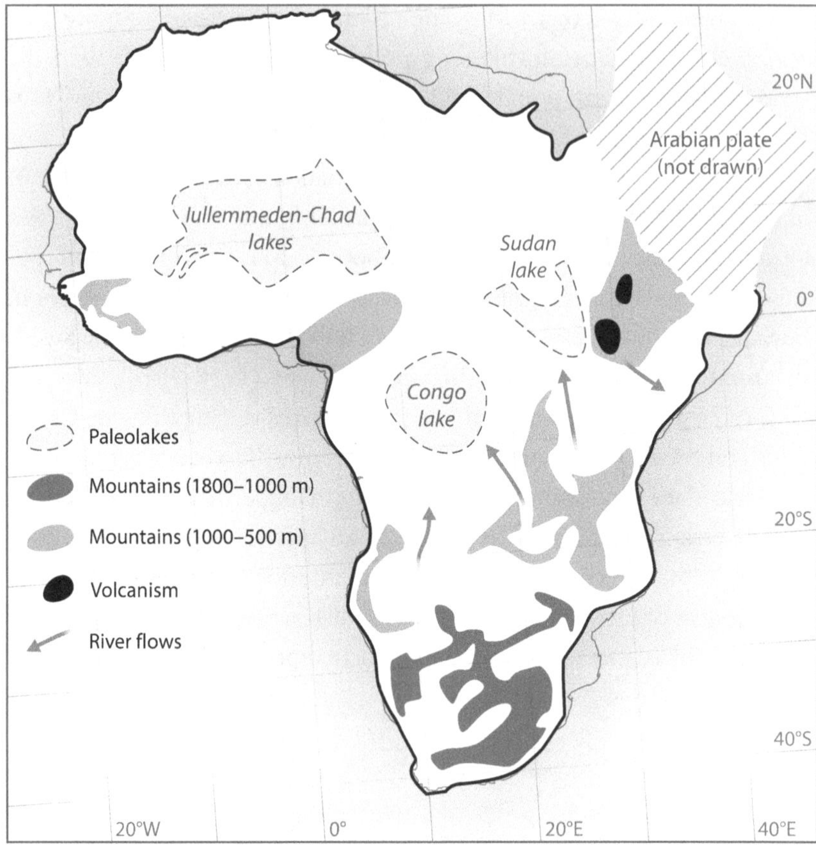

Figure 2.3. Paleogeographic map showing details of the paleotopography of Africa during the middle Eocene (48–41 Ma). The map shows the main alluvial basins (*white* areas) among mountains in Island Africa. The *arrows* indicate major river flows toward the center of the vast alluvial plains. Note the huge extension of lakes into the African interior during the Eocene. This map does not show detail on the Arabian plate, which was then connected to Africa in a single Arabo-African landmass. Redrawn by A. Lethiers from Couvreur et al. (2021, fig. 2B).

and floras. Instead, the ancient African mammal faunas display an overall homogeneity emphasizing the uniqueness of Africa in the picture of the Cretaceous and Paleogene faunal worlds.

The uniqueness of Africa stems from the isolation of the continent—Island Africa—and the related endemic evolution of its flora and fauna. Island Africa indeed gave birth to particular lineages and major evolutionary radiations. The African radiation resulted in an original adaptive and ecological mammalian world, including endemic insectivorous, carnivorous, and ungulate lineages that evolved in a convergent way to unrelated Eurasian and Laurasian mammals that had similar adaptations and ecology. There are, for example, antelope-like hyracoids, rhino-like ungulates (*Arsinoitherium*), carnivorous placentals (hyaenodonts), otter-like insectivores (otter shrews), mole-like insectivores (golden moles), saltatorial jerboa-like insectivores (sengis), rabbit-like rodents (springhares), termite eaters (aardvarks), and aquatic mammals (sea cows) all belonging to strictly African groups.

Yet, there is a paradox in the paleobiogeographic history of Island Africa. Study of Cretaceous and Paleogene African faunas reveals faunal relationships with Laurasian continents, in particular with Eurasia. There were interchanges (i.e., dispersals) of vertebrates across the Tethys Sea, which separated Africa and Eurasia. These "trans-Tethyan" dispersals enriched the indigenous African fauna. They were at the origin of several mammalian populations during the Paleogene (see "The Success of the African Story" in chapter 6). In the same way, rare but remarkable faunal interchanges also occurred between Island Africa and South America during the Paleogene. As far as we know, it was only dispersal to South America. These dispersals explain the African origin of New World monkeys (Platyrrhini) and of guinea pigs and other related rodents (the caviomorphs). These dispersals between the continental islands of Africa and South America were very occasional and took place in a still-mysterious way across the South Atlantic, but they most likely occurred on floating islands (plant rafts). Thus, the Cretaceous and Paleogene African faunas evolved in a generally insular context that was punctuated by several episodes of intercontinental faunal interchanges from the beginning of the history of Island Africa. The occasional interruption of African isolation by intercontinental dispersals is a feature of Africa's paleobiogeographic history when compared to other insular continents such as South America or Australia, which experienced a stronger, "splendid isolation," to use George Gaylord Simpson's now classic expression (Simpson, 1980).

The history of Island Africa ends with the closing of the Tethys Sea and the collision of Africa with Eurasia at the beginning of the Miocene, 23 Ma. It is the beginning of today's Old World faunal realm, which includes Africa, Eurasia, and India. The collision of Africa and Eurasia permitted great intercontinental dispersals and faunal interchanges between Africa, Eurasia, and India. This founder *geodispersal* event, which my colleague Jean-Claude Rage and I called the "Great Old World Interchange" (GOWI; Rage and Gheerbrant, 2020), is at the origin of the composition of the living African faunas, which include a mixture of ancient native lineages (elephants, flying squirrels, anthropoids) and late Eurasian arrivals (zebras, rhinos, lions, buffalo, and many others).

Paleoenvironments in Island Africa

Africa is the second-largest continent in the world (it covers about 20% of the world continental surface), and it was the largest part of the Gondwana supercontinent, which existed during the Mesozoic. Africa is one of the most tectonically stable continents: It is bounded by only passive tectonic margins, and it is structurally composed of a few large cratons (old and stable parts of the continental lithosphere) consisting of ancient and stable continental basements, with few mountain ranges, mainly those of the North African Atlas Mountains, the East African Rift (Kilimanjaro), and South Africa. The geography of the continent consists mainly of vast plains and alluvial basins such as those of the Congo, Niger, and Nile Rivers.

In the Paleogene, the Arabo-African island continent was in a more southerly paleogeographic position, with its widest part located at the level of the equator. Climatic conditions in Island Africa were relatively stable with a predominantly tropical climate. However, a significant climatic warming has been reported during the middle Cretaceous, close to the time of the separation of Africa and South America, with very high temperatures, especially during the "mid-Cretaceous hothouse" by 94–84 Ma (Huber et al., 2018; Russell and Paesler, 2003; Scotese, 2021). In particular, the Cenomanian–Turonian thermal maximum, at 94 Ma, was one of the warmest times of the Phanerozoic. It would explain the event called the Cretaceous Terrestrial Revolution, which affected the flora and fauna of the continental ecosystems of the middle Cretaceous, between 125 and 80 Ma. In Africa, it is thought to have caused major extinctions of the native fauna, and it may have favored the colonization and diversification of modern mammals in the continent (Rage and Gheerbrant, 2020).

The paleotopography of Island Africa during the Paleocene and Eocene was relatively flat, with no major relief and a low mean altitude close to sea level (Poblete et al., 2021) **(fig. 2.3)**. The few areas of high relief known at that time correspond to an ancient South African high plateau and to localized late Eocene mountains in North Africa (e.g., Tibesti and Hoggar). Vast tropical forests stretched across the great African lowlands. It was an Island Africa mostly covered by tropical forests in the interior and by mangroves on the coasts (including those of the interior trans-Saharan Sea in the Paleocene).

In the African Paleogene vegetation, the legume family was especially well developed, and palms were abundant, predominating in the lowland swamps. Some authors (e.g., Morley, 2000) reported indications of a latitudinal zonation of the vegetation until the late Eocene, and there is a provincialism of the flora in places, notably in the Saharan regions and in South Africa (with more temperate climate and Australian affinities). During the early Eocene there was a change in the vegetation of the lowlands with the appearance of a mosaic of dry and humid forests, and with the development of savannas (e.g., Central Africa). In the middle Eocene (45 Ma), arid regions and more open environments developed in places, notably in eastern (Tanzania), southern, and northern Africa (Egypt), where herbaceous and cycad vegetation grew. Open environments developed more markedly in the Oligocene.

The structural and climatic stability, the homogeneity of the geography and of the predominating tropical forest environment, and the ancient insular continental context all influenced the dynamics of mammal evolution and diversification in Island Africa cradle.

Paleogene climates in Island Africa

Our knowledge of the Late Cretaceous and the Paleogene climates comes mainly from the study of continental fossil flora and from the geochemistry of stable oxygen isotopes (^{18}O) in benthic and planktonic foraminiferal tests. In Africa the predominant climates during the mid-Cretaceous and the Paleogene were warm (average global temperatures 23°C) and humid, with a widely expanded tropical belt. It was mostly a hothouse world but with some variation over the time. The warm middle Cretaceous period became mild during the latest Cretaceous, and a short cooling event occurred at the KPg transition (an impact winter event). In the Paleocene and in the Eocene, the climates were again globally warm with little contrast (low seasonality) and with a low latitudinal thermal gradient. It was a period with a greenhouse effect and without ice ages. For context, the oxygen geochemistry of benthic foraminifera indicates deep ocean temperatures of 10°C–15°C for the Maastrichtian–Eocene period—compared to 1°C–2°C today.

The early Eocene was characterized by several thermal peaks, the most remarkable of which was that of the Paleocene–Eocene transition at 56 Ma. This is the so-called Paleocene–Eocene Thermal Maximum (PETM) (Gingerich, 2006). The overall African climate was hot and humid during the Eocene. This resulted in important deposits of bauxite and extensive precipitation of silica (e.g., in the Moroccan phosphate series). The middle Eocene paleoflora of Tanzania indicates high rainfall (640–780 mm per year) and the presence of forest vegetation (Jacobs and Herendeen, 2004). About 35 Ma, at the onset of the Oligocene, a major climate change is recorded with a global cooling. It was at this time that the ancient hothouse world was replaced by the icehouse world we know today. In Africa, this resulted in the development of arid zones in the north, east, and south of Africa. Aridity increased at the end of the Oligocene and the beginning of the Miocene.

A heterogeneous African fossil record with long gaps

Paleogene mammals of Island Africa remain poorly known: They are among the poorest known of all the continents. Even compared to the faunas of Gondwana, they are poorly documented, far behind those of South

America—although better known than those of Australia, Antarctica, and Madagascar. Moreover, the African fossil localities have an uneven geographic distribution, being essentially concentrated in North Africa **(fig. 2.4)**. They also have a disparate faunal diversity, which is notoriously low in many of them. Most African Paleogene faunas are poor or partially known, with the remarkable exception of those of the early Oligocene of the Fayum (Egypt), which were among the first discovered in Africa.

Thirty main fossil localities with *terrestrial* mammals have been discovered in the Arabo-African island for the Paleogene period, between 66 and 23 Ma. **Figure 2.5** shows the distribution in time of the most important localities. In all known African sites, 311 species in 216 genera have been described (author's 2024 inventory). Compared to, for example, the 900 species and 800 localities known at the same time in Europe according to the Paleobiology Database (https://paleobiodb.org), which probably provides underestimated numbers; for instance, the old inventory of Leduc (1996) has recorded at least 1,200 Paleogene species in Europe. To put this into a wider African perspective, 350 mammalian localities that include 800 genera have been discovered in Africa for the entire Cenozoic period (Van Couvering and Delson, 2020).

Significant fossil gaps darken several periods of evolution in Africa. Late Cretaceous and early Paleocene faunas are entirely unknown. There are also gaps in the Eocene fossil record **(fig. 2.5)**, especially for the middle Eocene between 41 and 38 Ma (Bartonian stage), during which, for example, proboscideans are known from a single specimen. Mammals from the late Oligocene (28–23 Ma) were unknown at the beginning of the 21st century. Since then, new fossil discoveries in the late Oligocene of Africa have followed in rapid succession, and as of 2024 some 30 species of this age are identified from eight African localities. However, there are still important groups, such as the aardvarks (tubulidentates), for which no Paleogene fossils have been found, although their phylogeny indicates an ancient African history.

Apart from a few notable exceptions, such as the Fayum (Egypt) and Dur at Talah (Libya), most African sites are of modest richness in fossils. They have yielded—most often after screening and washing several tons of sediment—fragmentary remains and essentially isolated teeth

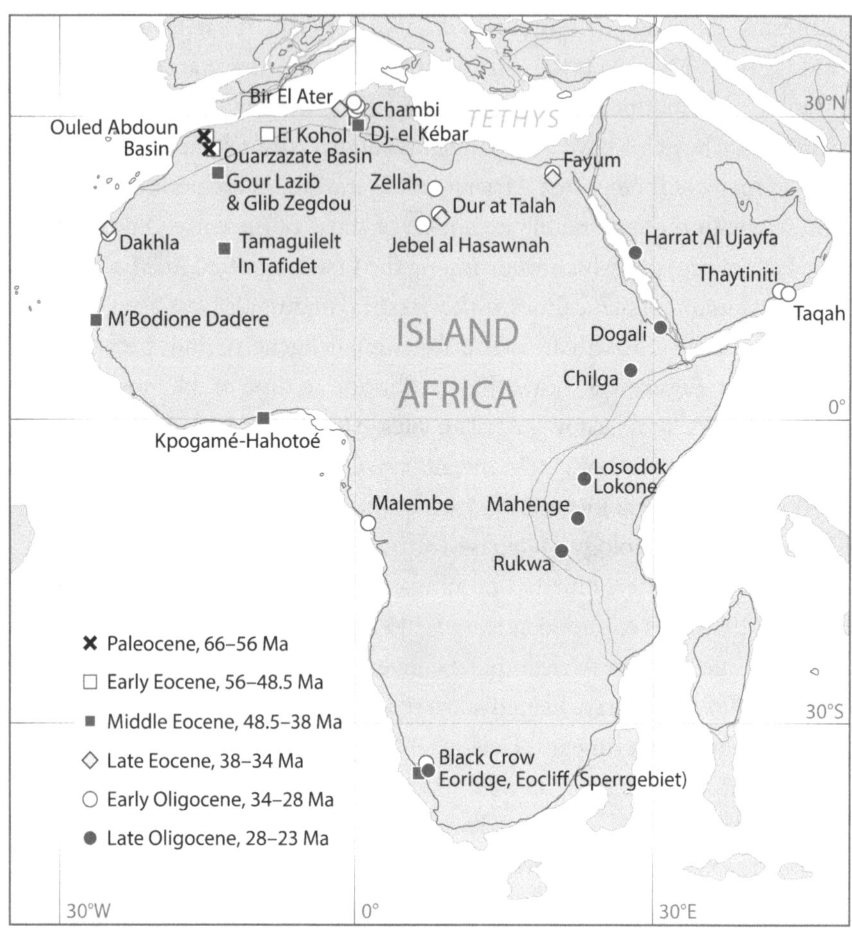

Figure 2.4. Main Paleogene (66–23 Ma) mammal localities discovered in Island Africa. Redrawn by A. Lethiers from a paleomap reconstructed for 34 Ma (early Oligocene) on the ODSN Plate Tectonic Reconstruction website, https://www.odsn.de/odsn/services/paleomap/paleomap.html.

(for instance, the Ouarzazate Basin, Chambi, Bir El Ater, Zellah, Taqah, and Thaytiniti sites). The site of Jebel al Hasawnah in Libya, which is still unexcavated, contains the only articulated skeletons of terrestrial mammals discovered in Island African from during the Paleogene (Gheerbrant et al., 2007). The immense fossiliferous outcrops of the remote Dur at Talah site also remain poorly explored and exploited.

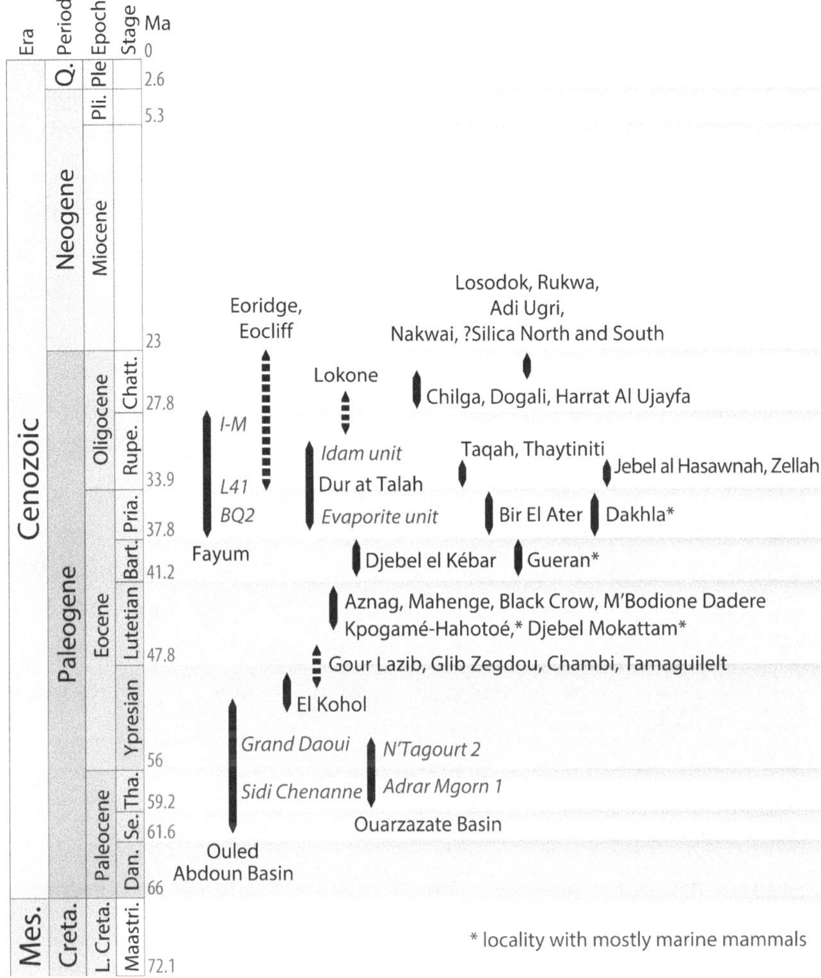

Figure 2.5. Stratigraphic distribution of the mammal localities known in the Paleogene of Africa (i.e., Island Africa). Note that the Namibian Sperrgebiet sites of Eoridge and Eocliff are considered here as, respectively, early Oligocene and late Oligocene, and that Silica North and Silica South sites are treated as either late Oligocene or early Miocene (see chapter 4, "Early Oligocene"). *Dotted lines* indicate the possible age of poorly dated sites, while *solid lines* indicate the age of a local fossil site or the stratigraphic extension covered by several fossil sites. Local sites belonging to the same large fossiliferous basin are shown in *italics*.

Only two main African Paleocene localities are known, both of which are in Morocco. They are the Ouarzazate Basin (Thanetian, ca. 57 Ma) and the Ouled Abdoun Phosphate Basin (Selandian, 61.5–59 Ma, and Thanetian, 59–56 Ma). Rare Ypresian sites (56–48 Ma) are known in Morocco (Ouled Abdoun and Ouarzazate Basins) and Algeria (El Kohol) only. The best-known Eocene faunas come from the Lutetian (48–41 Ma) of Algeria (Hamada du Dra: Gour Lazib and Glib Zegdou; 48 Ma), Tunisia (Chambi; 48 Ma), and Namibia (Sperrgebiet–Black Crow; ca. 43 Ma). Bartonian (41–38 Ma) African mammals are known from a single continental locality from Tunisia (Djebel el Kébar).

The last stage of the Eocene, the Priabonian (38–34 Ma), is fairly well documented in Libya (Dur at Talah), Morocco (Dakhla), Algeria (Bir El Ater), and Egypt (Fayum–Birket Qarun, Fayum-BQ2). The early Oligocene—that is, Rupelian (34–28 Ma)—is by far the best-represented period for the African Paleogene fauna, with primarily the famous Egyptian localities of the Fayum (90 local sites in the Jebel Qatrani Formation, of which 10 are rich and productive), and also those of Libya (Dur at Talah), and Oman (Taqah, Thaytiniti). There are nearly three times more species described from the Rupelian than the richest other Paleogene stages in Africa; this can be seen in **figure 2.6**, which shows the number of species by stage.

Our paleontological knowledge of the late and latest Oligocene (28–23 Ma) in Africa has progressed enormously with several new discoveries in eastern Africa, Tanzania, Kenya, Ethiopia, and Eritrea, and probably also in Namibia.

The paucity of the Cretaceous–Paleogene vertebrate fossil record in the vast Arabo-African continental plate is intriguing. The restricted quantity of known localities, their heterogeneous stratigraphic and geographic distribution, and the low diversity of most faunas are likely explained by gaps in our knowledge. We are dealing either with an insufficiency of field research (incomplete knowledge) or an absence of favorable fossiliferous localities (sedimentary gaps or biases). However, the African continent was explored as early as the beginning of the colonial period for a great deal of naturalist and mining research.

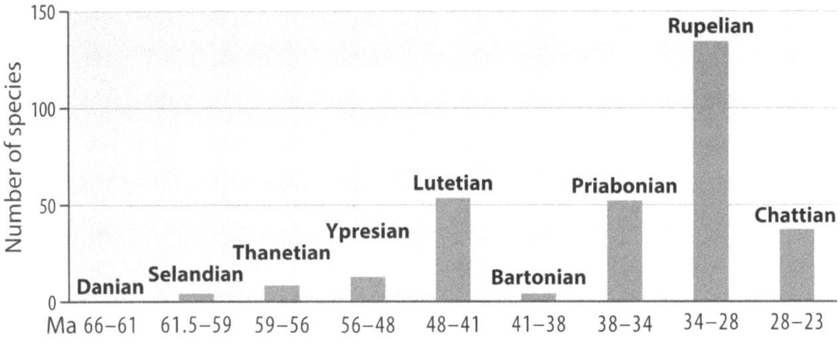

Figure 2.6. Number of known mammal species by stage during the Paleogene in Island Africa. As of 2024, 311 species of mammals have been identified from Island Africa, from a period lasting 43 million years, which corresponds to the first two-thirds of the Age of Mammals.

The structural stability of the continent prior to the Miocene might explain, in part, the relative scarcity of fossils (Coryndon and Savage, 1973): The geological layers have been relatively undisturbed with little uplift, which means moderate erosion and low sedimentation rates—in other words, there was less chance of preserving the skeletal remains of animals. This also implies that the ancient, or deep, fossiliferous levels are less exposed than in the structurally more mobile continents, such as Eurasia, where orogeny (mountain rise) brought the ancient layers up to the surface, notably those of the Cretaceous, which crop out widely and have yielded rich faunas. The regions of the Alpine belt, Mongolia, China, and central Asia are, for example, famous for their rich Mesozoic vertebrate localities. In addition, many of the African fossil mammal localities are of marine or near-coast facies (sedimentary environment).

The middle Eocene period is notably dominated by marine localities in Africa, and the Eocene generally represents a period of high sea levels compared to the Oligocene. The great continental shelf of the Arabian Peninsula and vast areas of the African continental margins were thus extensively inundated at this time, reducing areas for preservation of terrestrial faunas. There are very few known Paleogene continental fossil levels in the high peneplains of the great African interior, except for the Sahara. It is worth noting that, at the time, this was the area of the biotopes of immense tropical rainforests of the African hinterland.

For some paleontologists, the low diversity of Cretaceous and Paleogene faunas in Island Africa and the absence of several important vertebrate groups known elsewhere in Gondwana and Laurasia, such as multituberculates, are the results of a combination of the long African isolation, from 100 Ma to 23 Ma, and major Cretaceous extinction events, such as the Cretaceous Terrestrial Revolution (125–80 Ma) and the KPg crisis (66 Ma) (Rage, 1996; Rage and Gheerbrant, 2020). In any case, continental faunas of the Arabo-African island are certainly incompletely known. Compared to South America, this could be due to overall less-developed field research by local paleontological teams. Thus, there is still crucial field exploration work to be done to elucidate the evolution of mammals in Island Africa, particularly for the Late Cretaceous, the Paleocene, and the early and middle Eocene periods.

Poorly dated African fossil localities

The best-dated mammal localities on the Island Africa are of marine facies, represented by coastal or paralic sediments. These are the sites of the Ouarzazate Basin (Thanetian, Ypresian, Lutetian), the Ouled Abdoun Basin (Selandian, Thanetian, Ypresian), Zellah (Rupelian), Dakhla (Priabonian), Taqah (Rupelian), and Thaytiniti **(see fig. 2.4)**. The age of the Thaytiniti and Taqah sites is well supported by the presence of the benthic foraminiferan *Nummulites fichteli*, which is a marker species of the Rupelian stage (33.9–28.1 Ma). The age of the sites in the Ouarzazate Basin and the Ouled Abdoun Basin is determined from the associated shark faunas. Magnetostratigraphic data and recently obtained geochemical data (based on the study of stable isotopes of carbon and oxygen) have helped to constrain the age of these localities and to correlate them on a global scale. The ages of strictly continental localities, such as those of El Kohol, Chambi, Gour Lazib, Glib Zegdou, Tamaguilelt, Bir El Ater, Dur at Talah, Sperrgebiet, Chilga, and Lokone, are much more poorly constrained; their age is inferred essentially based on the evolutionary degree of mammals and the associated charophyte flora, in addition to the stratigraphic context.

Another difficulty for the dating of the Arabo-African Paleogene mammal localities is the endemism of their faunas. It hinders large-scale biostratigraphic correlations with the better-known and well-dated mammal faunas of other continents, such as Europe, North America, and Asia. African faunas are indeed represented by very different and unrelated animals. Correlation of the African continental sites requires global markers, notably geochemical and geochronological data. Geochemical data have only been recently obtained for some Paleocene and Ypresian sites in Morocco (Ouarzazate and Ouled Abdoun Basins). In the Ouled Abdoun, for instance, they indicate that the oldest-known African placental mammals are of Selandian age (ca. 60 Ma).

Evolution on the African island took place under very specific conditions, giving rise to a unique mammalian fauna. However, our current paleontological knowledge of early African mammals remains fragmented at several levels: geographic (uneven distribution of the fossil sites), temporal (sites of uncertain dating and gaps in the fossil record), regional (intra-African provincialism), and environmental (climate, flora, etc.). The paleontological exploration of Island Africa, although dating back to the 19th century, is far from complete.

CHAPTER 3

Mammalian Paleontology in Island Africa

Prior to 66 Ma, during Cretaceous times: An African terra incognita of origins

The fossil record of mammals in Africa at the time of the non-avian dinosaurs (Mesozoic) is one of the poorest in the world **(fig. 3.1)**. No stem or crown marsupial or placental mammals are known before the Cenozoic in Africa. The Late Cretaceous Period is a blank page for these mammals in Africa. During this period, extending from 100 Ma to 66 Ma, only one isolated mammal vertebra and a lower jaw of a possible gondwanathere are known from a site in Libya and Tanzania, respectively. Late Cretaceous mammals of Africa are a major focus of current paleontological field research.

In earlier times, during the Jurassic and Cretaceous, mammals known in Africa belonged to extinct archaic mammaliaform groups and to some stem groups that are placed at the base of the therian mammals, which include modern (extant) mammals such as placentals and marsupials. The most important discoveries have been made by paleontologists from the Muséum National d'Histoire Naturelle in Jurassic and Jurassic–Cretaceous transitional levels found in the Anoual region of Morocco **(plate 1 upper)**. The first discoveries were made more than 30 years ago by Denise Sigogneau-Russell and colleagues (1988) in the Jurassic–Cretaceous transitional levels, and important finds were later reported in a new Middle

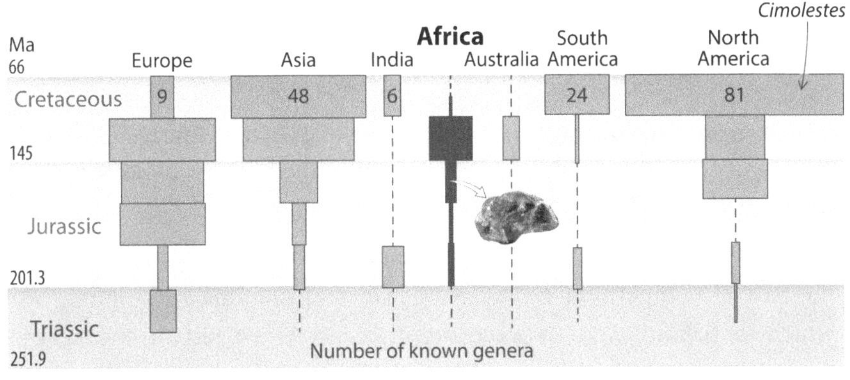

Figure 3.1. Mesozoic mammal fossil record of Africa compared with the records of other continents in number of genera. The width of *horizontal bars* shown for each major period is proportional to the number of genera; number of genera is in the bars only for the Late Cretaceous, except Africa, for which we know only one undetermined small vertebra from the Cenomanian (100–94 Ma) and one possible gondwanathere of uncertain Turonian–Campanian age (91–72 Ma). Note that there are no known fossils in Africa from the end of the Cretaceous, a critical period in the evolution of modern mammals. Jurassic and Cretaceous mammals from Africa are known especially in the Anoual sites (Morocco), which have yielded, among other groups, a diversified assemblage of dryolestoids (*inset*, photo of a lower molar). *Cimolestes* is one of the best-known Late Cretaceous eutherians from North America. Redrawn from Kielan-Jaworowska et al. (2004) and updated based on the Paleobiology Database, https://paleobiodb.org.

Jurassic site by Haddoumi and colleagues (2016) and in the Jurassic–Cretaceous levels by Lasseron and colleagues (2019). The review and detailed study of the Anoual fauna in the recent PhD thesis of Maxime Lasseron (2020) identified some 20 mammal species of six major groups (Haramiyida, Eutriconodonta, "Symmetrodonta" or basal Cladotheria, Dryolestoidea, Zatheria, and Boreosphenida). As a result the Anoual sites have yielded the richest Mesozoic mammal association discovered in Africa and even in Gondwana. They yielded, in particular, some of the oldest tribosphenic mammals known in the world—that is, the earliest-known ancestral therian mammals, which invented mastication and are placed at the root of both the extant marsupials and placentals. Lasseron's thesis highlighted that, despite a very poor fossil record, the African Mesozoic is characterized by a high frequency of stem and early therians (cladotheres such as dryolestoids, zatherians, and boreosphenids).

From 66 Ma to 23 Ma, the Early Cenozoic (Paleogene): Fossil discoveries in the African cradle

The African Cenozoic fossil record contains no archaic lineage of Mesozoic mammals, unlike other Gondwanan continents such as South America. Island Africa was an open evolutionary space in the early Cenozoic for modern mammals. The new ecological opportunities it provided and the lack of competition explain why placental mammals (the group to which we belong) were so successful in Island Africa early in the Cenozoic. The marsupials ("pouched mammals") arrived in Africa much later than placentals, in the Oligocene by 34 Ma, and they remained discrete during their evolution in Island Africa with a few species of a stem Laurasian family, the Herpetotheriidae, which rapidly became extinct, their last-known occurrence being in the middle Miocene (*Morotodon aenigmaticus*). The most successful African placental groups are the afrotherians (the word "Afrotheria" literally means "African beasts"); among them are the endemic ungulates called paenungulates ("near ungulates"), such as hyraxes, elephants, and sea cows, that rapidly colonized the herbivorous niches of Island Africa in an opportunistic manner. Also thriving, and being among the most abundant in late Paleogene localities, were the rodents and the primates, such as anthropoids. True carnivores (Carnivora) are among the many Laurasian mammals typically absent from Island Africa, an absence related to the geographic isolation of the continent. Instead, an original group of carnivorous placentals—now extinct—emerged and diversified in Island Africa: the hyaenodonts. All these mammals—paenungulates, primates, rodents, and hyaenodonts—were abundant and predominant in the Arabo-African localities of Oligocene age.

At the very beginning of the Cenozoic, during the Paleocene, mammal diversities remained low, and the relationships and identity of known groups at this time remain obscure due to the scarcity and fragmentary nature of the fossils discovered. The main Paleocene lineages are paenungulates and their stem groups, hyaenodonts, and small generalized insectivore-like species corresponding to ancestral or early placentals of still-unresolved precise relationships. Some of these are likely related to the root of extant afrotherian insectivores (tenrecs, otter shrews, golden moles) and hyaenodonts. Another interesting extinct group of (probably arboreal) small insectivores belongs to the enigmatic family Adapisoriculidae, also known from Europe and India. Its relationship is much debated. In the most recent phylogenetic studies, this family has been placed outside the modern placentals—that is, among the stem groups (basal eutherians) of all living placentals.

Fossil discoveries of Paleogene mammals in Africa date back about 150 years. Pioneering field explorations and discoveries were made in North Africa, with about 20 fossil sites found in Algeria, Tunisia, Libya, and Morocco. These include some of the most famous fossil localities, such as those of the Fayum Depression, in Egypt. English, French, German, and American paleontologists were involved in these early discoveries.

Late 19th century: Pioneering discoveries

The first Paleogene mammals found in Africa were marine. They were early sea cows (sirenians) and then early whales (cetaceans) from the middle and late Eocene of Egypt (marine formations of Mokattam and Qasr El Sagha). We owe their discovery to the workers of the building-stone quarries of the Cairo region (Djebel Mokattam) and to the field prospecting and geological studies of the German naturalist G. Schweinfurth from 1877 to 1886 in the Birket Qarun region. These early paleontological discoveries prompted the first publications on the Paleogene mammals from Africa at the end of the 19th century (Dames, 1883, 1894; Filhol, 1878; Owen, 1875).

In 1891, Albert Gaudry (1827–1908), the first professor of paleontology at the Muséum National d'Histoire Naturelle in Paris, described a tooth of a proboscidean named *Mastodon turicensis* from the Tunisian site of Khenchela, at that time presumed to be of Miocene age (Gaudry, 1891). This tooth was later recognized as belonging to the early proboscidean genus *Moeritherium* (Schlesinger, 1913; Tassy, 1981). *Moeritherium* is thus the first Paleogene terrestrial mammal discovered in Africa.

Early 20th century: First discovery of land mammals in the Fayum

The early 20th century saw the first paleontological excavations in Egypt. They were focused on large mammals in the fossil localities of the famous Fayum Depression (Qasr El Sagha and Jebel Qatrani Formations). These early excavations were organized by the paleontologists Hugh J. Beadnell (1874–1944) and Charles Andrews (1866–1924) of the British Museum of Natural History (the Natural History Museum, or NHM, today) from 1894 to 1907. These pioneering paleontologists in Africa made the first announcements about the amazing large terrestrial mammals of the early Oligocene Jebel Qatrani Formation **(plate 1 lower)**. Their discoveries included the early elephantiform proboscideans, embrithopods, hyracoids, anthracotheres, and hyaenodonts. Among them, the proboscidean *Palaeomastodon* was the first-reported terrestrial mammal from the Fayum (Andrews, 1901a, 1901b, 1901c).

Early research in the Fayum was crowned by the publication of the big monographs by Beadnell (1905) on the geology of the sites and by Andrews (1906) on the paleontology of the vertebrates (including mammals), which have become classic references on the Fayum fossil sites. Other early field research was carried out in the early 1900s by the German paleontologists Ernst R. Stromer (1871–1952) and Max Blanckenhorn (1861–1947), mostly in the underlying marine late Eocene levels (Qasr El Sagha Formation). This research provided new data on the archaeocetes, the extinct toothed whales (Blanckenhorn, 1903; Stromer, 1903).

This pioneering research in the Fayum was followed in 1907 by the American Museum of Natural History (AMNH) expedition lead by Henry F. Osborn (1857–1935), who was especially interested in searching for early elephants (proboscideans). This expedition, assisted by paleontologists Walter Granger, George Olson, and Richard Markgraf, made major discoveries. The Austrian amateur paleontologist Richard Markgraf (1869–1916) was the first to be interested in the search for small mammals (the so-called microfauna), and he discovered the first early anthropoids (primates) in the Fayum with jaws of the genus *Apidium*

(Osborn, 1908). *Apidium* was first identified as a bovid ungulate ("Apis" was a sacred bull of ancient Egypt) before being recognized as a primate thanks to new discoveries (Schlosser, 1910, 1911). Markgraf's field surveys led to the discovery of several new groups in the Jebel Qatrani Formation, all of small mammals: rodents, bats (chiropterans), elephant shrews (macroscelideans), hyaenodont carnivorous mammals, and the strange, extinct ptolemaiids (Osborn, 1908, 1909). Markgraf continued his fieldwork research for the Royal Natural History Office in Stuttgart, Germany, until 1911 with new fossil collections of small mammals. The German paleontologist Max Schlosser (1854–1932) studied these fossils and identified several new anthropoid primates (Schlosser, 1911). He also published new information about the large mammals. After the Second World War, in 1947, another American paleontological expedition to Fayum was led by Robert Denison and the South African paleontologist H. Basil Cooke (1915–2018) of the University of California.

Second part of the 20th century: The quest for earliest African placentals

The first discoveries of terrestrial mammals in Africa outside the Fayum were made from the 1950s onward by the French paleontologists Camille Arambourg (1885–1969), René Lavocat (1909–2007), and Philippe Magnier. The most important new fossil sites were those of Dur at Talah **(plate 2 upper)** and Zellah, in Libya, which are of similar age to those of the Fayum. These sites are rich but have remained little studied for a long time. Excavations in the Fayum were reopened from 1961 to 1968 by the team led by the pioneering paleoprimatologist Elwyn L. Simons (1930–2016), from Yale and Duke Universities (Simons et al., 1968). The team focused on the early Oligocene Jebel Qatrani Formation, and especially on the upper levels (the "upper fossil wood zone" or "upper sequence") that were still poorly known. Field surface research on the fossiliferous bed concentrated on small mammals, such as rodents (Wood, 1968) and primates, for which the team collected more complete and informative fossils. This led to the discovery of the oldest-known and famous catarrhine *Aegyptopithecus* (Simons, 1965).

The discoveries of early Cenozoic mammals in Africa, from Paleocene and early Eocene beds, were made in North Africa in the 1970s and 1980s by French paleontological teams (University of Montpellier, University of Paris 6, and Muséum National d'Histoire Naturelle). They were found in Morocco, in the Ouarzazate Basin (Adrar Mgorn 1, **plate 8**; N'Tagourt 2; Cappetta et al., 1978), and in Algeria with the early Eocene site of El Kohol **(plate 3)**. These sites yielded the oldest fossils of several important lineages, such as stem elephants (proboscideans), hyracoids, elephant shrews (macroscelideans), "insectivores" in the broad sense, primates, hyaenodonts, bats (chiropterans), and rodents. Major discoveries include the oldest primate *Altiatlasius* (Sigé et al., 1990) and the early proboscidean *Numidotherium* (Mahboubi et al., 1984).

The first middle Eocene mammal faunas from Africa were discovered in 1975 in Algeria (Hamada du Draa, Gour Lazib, and Glib Zegdou; Gevin et al., 1975) and in 1985 in Tunisia (Chambi; Hartenberger et al., 1985). Late Eocene continental sites were found at the same time in Algeria (Bir El Ater; Coiffait et al., 1984). Bir El Ater has yielded one of the oldest anthropoid primates known in Africa (De Bonis et al., 1988). Soon after, the first discoveries of Paleogene mammals in the Arabian Peninsula were made by Herbert Thomas (Collège de France), Jack Roger (Bureau de Recherches Géologiques et Minières), and Sevket Sen (Centre National de la Recherche Scientifique) in the Taqah and Thaytiniti sites of Oman (Thomas et al., 1988; Thomas et al., 1989) **(plate 2 lower and plate 4)**. The fossiliferous levels from these sites are of the same early Oligocene age as lower levels of the Jebel Qatrani Formation of the Fayum. The Omani sites yielded a great diversity of primates, including the first lorisiforms (Gheerbrant et al., 1993) but also including marsupials, chiropterans, hyaenodonts, rodents, and hyracoids. The paleontologist Martin Pickford from the Collège de France recognized the early Oligocene age of the Angolan mammal fauna of Malembe (Pickford, 1986, 1987), which was initially reported and correlated to the Miocene by Dartevelle (1935). As a result, this was the first Paleogene locality discovered in sub-Saharan Africa.

Field work in the Fayum was interrupted for about 10 years because of hostilities between Egypt and Israel. Field work resumed in 1977 with the American Duke University expeditions led by E. L. Simons. Field research of Simons's team led to the first discovery of marsupials (Bown and Simons, 1984) and pangolins (Gebo and Rasmussen, 1985), and to important new fossils of primates and ptolemaiids (Bown and Simons, 1987). This research also led to the discovery, in 1983, by T. M. Bown (US Geological Survey) of the important locality L41 (Rasmussen and Simons, 1991; Simons et al., 1998), which was excavated for nearly 25 years by the team of Erik R. Seiffert (Duke University and University of Southern California), a former student of E. L. Simons. This rich L41 site has yielded well-preserved fossils, including skulls, of small mammals belonging to 35 species of anthropoid and strepsirrhine primates, macroscelideans, hyracoids, rodents, hyaenodonts, and tenrecoid insectivores. From 1983, new research on the marine mammals of Djebel Mokattam, Wadi Hitan, and Fayum was undertaken by Philip D. Gingerich of the University of Michigan (Gingerich, 2024).

The last important discoveries of the end of the millennium were made in 1996 by the author of this book in the Ouled Abdoun Phosphate Basin, previously famous for its extraordinarily rich marine fauna comprising sharks, fishes, crocodilians, turtles, and giant extinct marine reptiles (mosasaurs, plesiosaurs). Gheerbrant et al. (1996, 1998) reported early Eocene fossils of one of the earliest-known proboscideans, *Phosphatherium*.

Third millennium (2001–2020): Lifting the veil on last endemic African fauna

The first 20 years of the new millennium were a period of intense paleontological field survey with numerous fossil discoveries, including from the previously blank periods of the late Oligocene, the middle Eocene, and the Paleocene. In recent decades there has been great international scientific emulation and competition in field research in Africa, in relation to major questions on evolution, especially on the phylogeny of placentals and on the origin of anthropoid primates. This led to the description of about 150 new species in more than 100 systematic papers. This is equivalent of all previous fossil discoveries over 130 years! Unfortunately, this golden age of field paleontology in Africa is over. Field prospecting and excavations have become difficult with the deterioration of the international geopolitical situation, particularly in Africa. Several key fossil sites have become inaccessible—for example in Libya, Tunisia, Mali, and Egypt, including those in the Fayum. Field research has also been hampered by funding problems in an equally difficult economic period.

Paleocene and early Eocene. The development of field research by the author in the Ouled Abdoun phosphate deposits (Morocco) has led to the discovery of the oldest African placentals, notably in Paleocene levels of the Selandian dated at about 60 Ma **(plate 5)**. These middle Paleocene mammal levels, unknown elsewhere in Africa, have yielded the first remains of African "condylarths" (Gheerbrant et al., 2001)—that is, stem paenungulates—and the earliest-known proboscidean *Eritherium* (Gheerbrant, 2009). The Ypresian Ouled Abdoun phosphate levels yielded additional well-preserved fossils, including those of a large proboscidean (Gheerbrant et al., 2002) and the earliest embrithopods (Gheerbrant et al., 2018, 2021).

Middle Eocene. The resumption of the excavations in the middle Eocene localities of Algeria (Gour Lazib and Glib Zegdou, Hamada du Draa) **(plate 6)** and Tunisia (Chambi) by the paleontologists Rodolphe Tabuce (University of Montpellier II) and Mohamed Mahboubi (University of Oran) has led to the discovery of a variety of endemic groups, including a remarkable diversity of hyracoids (Tabuce et al., 2007), the oldest African strepsirrhine primates (Marivaux et al., 2013; Tabuce et al., 2009), elephant shrews (Tabuce, 2018; Tabuce et al., 2001), bats (Ravel et al., 2011, 2016), and hyaenodonts (Solé et al., 2014, 2015, 2016). The first discoveries of Eocene mammals in sub-Saharan Africa, apart from Dartevelle's "Miocene" Angolan sites redated by Pickford, were made in the first decade of the 2000s. In 2003, Gregg Gunnell (1954–2017, Michigan and Duke Universities) and his team reported from Tanzania the partial skeleton, an exceptional fossil, of an endemic bat in the Lutetian sediments of the Mahenge crater lake (Gunnell et al., 2003). In 2008, Pickford published major new discoveries in Namibia in the diamond-bearing area Sperrgebiet (literally "forbidden zone"), particularly at Black Crow (Pickford et al., 2008) **(plate 7 upper)**. These sites include richly fossiliferous lacustrine limestones levels that are still being excavated. They have yielded a variety of small mammals associated with a few large species: afrotherian insectivores related to tenrecs, otter shrews and golden moles, and rodents, bats, lorisiform and adapiform primates, a fruit bat, hyracoids, and an embrithopod. Some of the mammal Namibian sites (Silica, Eoridge, Eocliff) found by the Pickford team are of debated age. The Tunisian site of Djebel el Kébar, discovered in 2014 by the team of Laurent Marivaux (University of Montpellier II), is the only one known from the Bartonian stage in Africa. It was dated at 39.5 Ma using the potassium–argon (K–Ar) radiometric method. For the moment, only a few teeth belonging to two species are described, a primate probably basal to the African anthropoids (Marivaux et al., "Morphological Intermediate," 2014) and a hystricognath rodent.

Late Eocene and early Oligocene. Forty years after the pioneering work of Robert J. G. Savage (1927–1998), the resumption of paleontological survey in Libya by the team of Jean-Jacques Jaeger (University of Poitiers) and Pauline Coster and Chris Beard (both of Kansas University) in the late Eocene of Dur at Talah (Jaeger et al., "Late Middle Eocene," 2010) and in the early Oligocene of Zellah **(plate 7 lower)** (Coster et al., 2015) yielded a significant mammal microfauna (Coster et al., 2012, 2015; Jaeger et al., "New Rodent Assemblages," 2010, 2012) including rodents, elephant shrews (Tabuce et al., 2012), anthropoid and adapiform primates (Jaeger et al., "Late Middle Eocene," 2010), and hyaenodonts (Grohé et al., 2012; Mattingly et al., 2020). Apart from fossils found in Mahenge (Tanzania), the only known articulated skeletons of Paleogene mammals in Africa were found in Libya, in the early Oligocene lacustrine levels of Jebel al Hasawnah. They belong to a hyracoid species **(plate 14)** previously described in the Fayum (Gheerbrant et al.,

2007, 2012). Erik R. Seiffert has continued the excavations in the Fayum previously undertaken by his thesis adviser E. L. Simons. His team has collected many new mammals, in particular from the prolific locality L41 (latest Eocene or earliest Oligocene, ca. 34 Ma) and in the older one of BQ2 (late Eocene). Noteworthy new finds are lorisiform and adapiform primates, rodents, hyaenodonts, hyracoids, the oldest African catarrhine primates (BQ2 site) comparable to those of Bir El Ater, and insectivores possibly ancestral to tenrecoids (the group of potamogales, tenrecs, golden moles).

Important new discoveries of marine mammals, in particular sea cows, were also made in the late Eocene levels of the "Valley of the Whales" of Wadi Hitan (Egypt) by the team of Gingerich and Zalmout (Gingerich, 2024; Zalmout and Gingerich, 2012). Similarly, new sites with marine mammals (especially toothed whales, archaeocetes), sometimes associated with remains of terrestrial species, were discovered in the Moroccan Saharan localities of Dakhla (early Oligocene and late Eocene) and Gueran (middle Eocene) by the team of Adnet, Marivaux, and Benammi and the team of Zouhri (University of Casablanca) and Gingerich (Adnet et al., 2010; Haddoumi et al., 2016; Zouhri et al., 2021). In 2012 the team led by Hautier (University of Montpellier) also uncovered marine mammals in phosphate levels of the Lutetian of Senegal (Hautier et al., 2012). These marine levels contain one of the oldest-known sea cows.

Late and latest Oligocene. A remarkable discovery is that of the first-known late Oligocene fossil sites in Africa. This discovery was made in 2003 by Kappelman (University of Texas) and Sanders (Michigan University) with the Ethiopian site of Chilga (28–27 Ma), which yielded large African ungulates (paenungulates) (Kappelman et al., 2003). These are embrithopods (*Arsinoitherium*) and proboscideans (palaeomastodonts) known in the Fayum but represented in Chilga by larger species (for example, the species *Arsinoitherium giganteum* of elephantine size). Some other mammals, unknown in the Fayum and more typical of the Neogene, are present at Chilga, including the first-reported Paleogene deinothere and a close relative of the elephantoids (gomphothere group). Other important discoveries were made mainly in the eastern part of Africa. In Eritrea, at the Dogali site, Jeheskel Shoshani (1943–2008, Wayne State University) described in 2006 a proboscidean intermediate between the early elephantiforms from the Paleogene and the modern elephantoids from the Neogene (Shoshani et al., 2006). In Tanzania, the site of Rukwa discovered by the team of Nancy Stevens (Ohio University) yielded, among other micromammals, an anthropoid primate close to the ancestry of the apes (hominoids), and in Arabia (Harrat Al Ujayfa site, 29–28 Ma) the team of Zalmout (Michigan University) and Gingerich described in 2010 a slightly more basal anthropoid, related to the common ancestry of the Old Word monkeys and apes (Zalmout et al., 2010). In Kenya, several late Oligocene sites (Lokone, Losodok) have yielded a more evolved fauna than that of the Fayum, including primates, hyracoids, proboscideans, hyaenodonts, and rodents (Ducrocq et al., 2010; Rasmussen and Gutierrez, 2009). This fauna is characterized by an intermediate composition between the ancient endemic Paleogene fauna and the more modern faunas of the Miocene, with, for example, the presence of a viverrid carnivore **(see fig. 4.31)** that foreshadows the Miocene Eurasian dispersal wave and the end of the African isolation. The localities of Adi Ugri and Mai Gobro reported by Abbate et al. (2014) in Eritrea are thought to be the most recent of the Paleogene of Africa (i.e., of terminal Oligocene age). They have yielded fragmentary fossils of modern proboscideans that heralds Neogene groups such as gomphotheres and deinotheres.

Paleontological knowledge of the paleobiodiversity of Island Africa mammals has progressed enormously with the development of fieldwork since the first discoveries about 150 years ago. However, our knowledge is still largely based on a few reference fossil sites from the early Oligocene of North Africa, such as those at Fayum, which have yielded the richest mammal assemblages known in Africa after over a century of exploration and excavation. Recent fossil discoveries have, however, begun to lift the veil on the earliest (Paleocene) and most recent (late Oligocene) periods in the history of mammals in Island Africa, and on regions long devoid of fossils, such as East Africa, sub-Saharan Africa, Namibia, and the Arabian Peninsula. The current state of paleontology allows the broad outlines of the evolutionary history and succession of mammals in Island Africa to be traced over much of the Paleogene, from the middle Paleocene (60 Ma) to the end of the Oligocene (23 Ma).

Plate 1. Upper. The Anoual microvertebrate locality, Morocco, Jurassic–Cretaceous transition (ca. 130 Ma). This site has yielded one of the most diversified microvertebrate assemblages from the Mesozoic of Gondwana—in particular, some 16 species of micromammals. **Lower.** The Fayum mammal site of Quarry M (*foreground*) in Egypt, Jebel Qatrani Formation (ca. 28 Ma). The flat hill in the background is Tel Richard Markgraf. Photograph taken during 2005 field expedition lead by Elwyn L. Simons and Erik R. Seiffert. The Fayum mammal sites are the richest known in Island Africa. They have yielded nearly half of all known Paleogene mammal species in Africa. Photo by E. Gheerbrant (**upper**). Photo courtesy of Mark E. Mathison (**lower**).

Plate 2. Upper. The Dur at Talah mammal localities, Libya, late Eocene (36–34 Ma) and early Oligocene (34–30 Ma). The fossiliferous outcrops of the Dur at Talah locality are exposed along a remote and immense cliff that extends for over 150 kilometers and remains poorly explored. **Lower.** The mammal locality of Taqah, Oman, early Oligocene (34 Ma). The Taqah site has yielded a diversity of small endemic African mammals (micromammals) that lived in a tropical rainforest near a shallow coastal marsh, along the southern coast of the Arabian Peninsula. The site includes a remarkable diversity of stem (parapithecoids) and modern (early catarrhines) anthropoids that are also found in the Fayum faunas. The mammal fossils (mostly teeth) from the Taqah site were recovered after screening-washing 17 metric tons of fossiliferous sediments. Photo courtesy of L. Marivaux (**upper**). Photo by E. Gheerbrant (**lower**).

Plate 3. The early Eocene El Kohol, Algeria, mammal site (ca. 51 Ma) that has yielded an exceptionally rich fossil sample of the early proboscidean *Numidotherium koholense*, associated to some other African mammals. **Upper.** General view of the site (green fossiliferous levels in the foreground). **Lower.** Detail of fossil jaws of *Numidotherium koholense* being excavated at the El Kohol site. Photos courtesy of R. Tabuce.

Plate 4. The 1992 paleontological camp at the Thaytiniti (Dhofar, Oman) mammal site (from 34 Ma) of the excavation team led by Herbert Thomas (Collège de France), located just above the fossiliferous level (light-colored beds). Photo by E. Gheerbrant.

Plate 5. The Ouled Abdoun phosphate quarries (Daoui sites, Morocco) yielding Paleocene and early Eocene (60–53 Ma) mammals. **Upper.** Quarries in the Grand Daoui area where the holotype of *Phosphatherium escuilliei* was found; mammal fossiliferous level shown (*arrow*). **Lower.** Excavations in Sidi Chennane quarries in search of mammals and other vertebrates in the Paleocene levels. The earliest-known elephant relatives (*Eritherium*) and the stem groups at the root of all the African ungulates (paenungulates) were discovered in the Sidi Chennane quarries. Photos by E. Gheerbrant.

Plate 6. The Glib Zegdou mammal localities (Hamada du Draa, Algeria) of the early or middle Eocene (ca. 48 Ma). **Upper.** Overall view of the site, including the fossiliferous levels (*arrow*). **Lower.** Detail of the excavation in the mammal levels by the paleontological team of Rodolphe Tabuce from University of Montpellier. Excavations at the site have uncovered well-preserved mammal fossil remains, in particular a remarkable variety of early hyracoids. Photos courtesy of R. Tabuce.

Plate 7. Upper. The Black Crow mammal fossiliferous site, middle Eocene of Namibia (ca. 43 Ma). The fossiliferous levels consist of lacustrine limestones. The *yellow star* indicates the place where the distal part of the skull of the embrithopod *Namatherium blackcrowense* was found, recently described by Gheerbrant et al. (2025). The holotype of *Namatherium blackcrowense*, a maxilla, was found 200 meters away. The *red star* corresponds to the type locality of the hyracoid of *Namahyrax corvus* and the site of fossils of other mammals (primates, rodents). **Lower.** The Libyan Zellah micromammal locality from the early Oligocene (ca. 34 Ma). Micromammals such as rodents and primates were recovered at this site by screening. Photo courtesy of M. Pickford (**upper**). Photo courtesy of K. C. Beard (**lower**).

Plate 8. The late Paleocene Adrar Mgorn fauna (ca. 57 Ma), located in the Ouarzazate Basin in southern Morocco, was discovered in fossiliferous limestones located at the top (*arrow*) of a hill called the Gara de Tinerhir, pictured here. This site has yielded a diversified microvertebrate fauna, including about 100 species of bony fishes, sharks, frogs, salamanders, lizards, snakes, turtles, crocodiles, and mammals (25 species). The recovered mammal fauna includes only small species ("micromammals") belonging to stem and more evolved placentals, including one of the first-known true primates, *Altiatlasius*. The Gara de Tinerhir hill seen here is around one kilometer long. Photo by E. Gheerbrant.

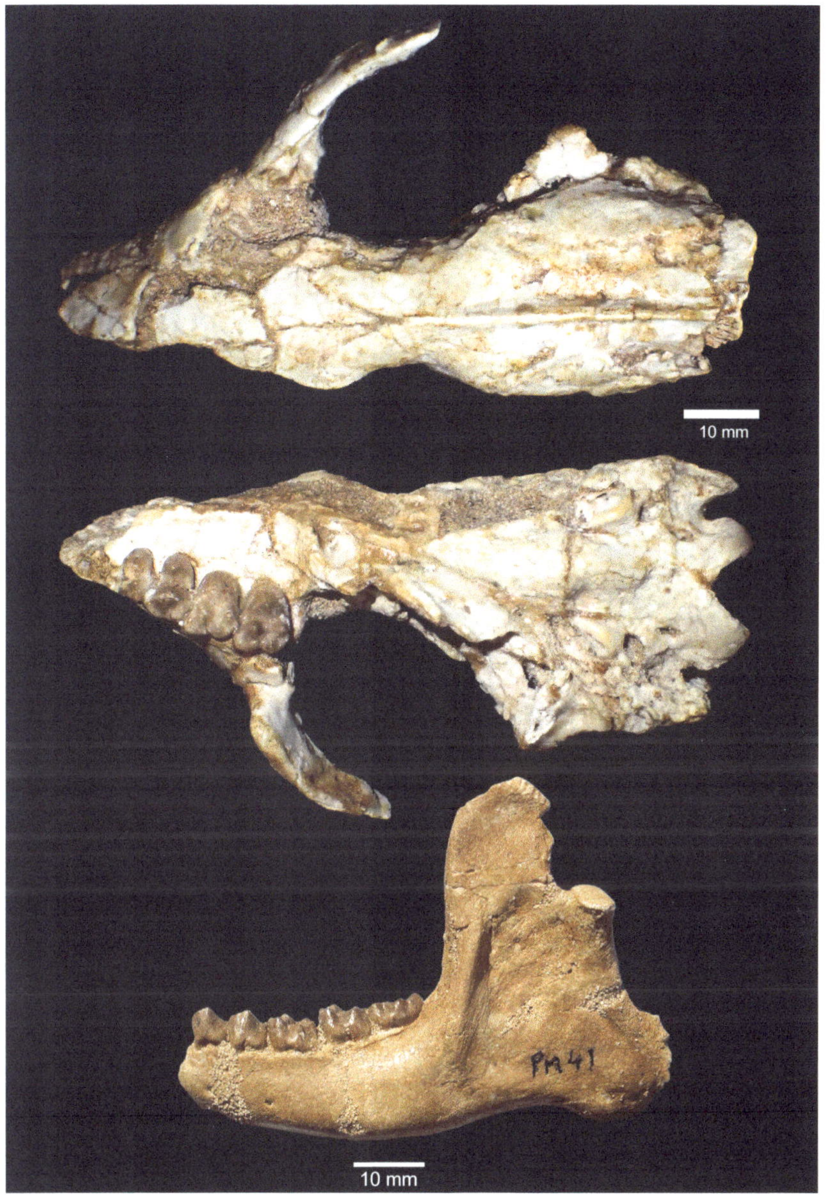

Plate 9. The skull (*top and middle*) of and lower jaw (*bottom*) of *Ocepeia daouiensis*, another stem paenungulate found with *Abdounodus* in the middle Paleocene (60 Ma) marine fossiliferous phosphate beds of the Ouled Abdoun Basin, Morocco. This *Ocepeia daouiensis* skull is the earliest-known skull of an afrotherian mammal. Photo by E. Gheerbrant.

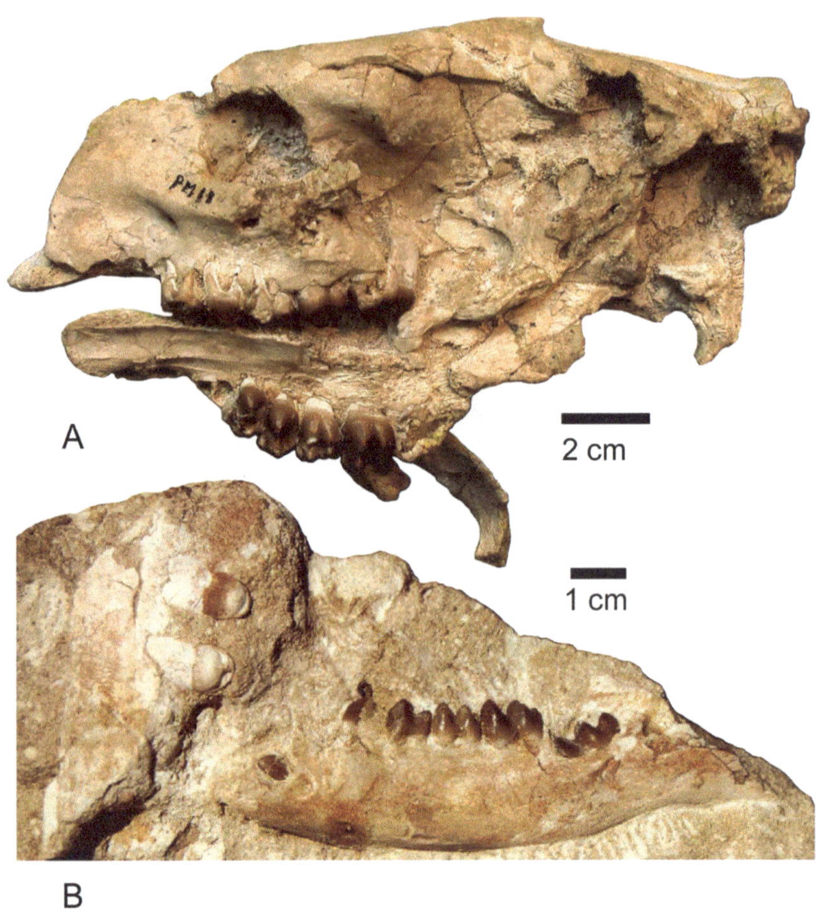

Plate 10. The early elephant relative (proboscidean) *Phosphatherium escuilliei* from the early Eocene Ouled Abdoun Phosphate Basin, Morocco (ca. 56 Ma). **A.** The best-preserved skull known in the species. **B.** A partial mandible still embedded in phosphate matrix. *Phosphatherium* was a basal and small proboscidean (body mass of 15 kg; shoulder height of 30 cm). Photos by E. Gheerbrant.

Plate 11. Skull and mandible of the stem proboscidean *Numidotherium koholense* from the early Eocene of El Kohol, Algeria (ca. 51 Ma). Many well-preserved fossils of *Numidotherium* have been found, documenting most of its skeleton. *Numidotherium* is the best known of the first relatives of elephants (proboscideans). *Numidotherium* was the size of a tapir (body mass 250–300 kg; shoulder height 1 m). From Noubhani et al. (2008, fig. 2); photos courtesy of R. Tabuce.

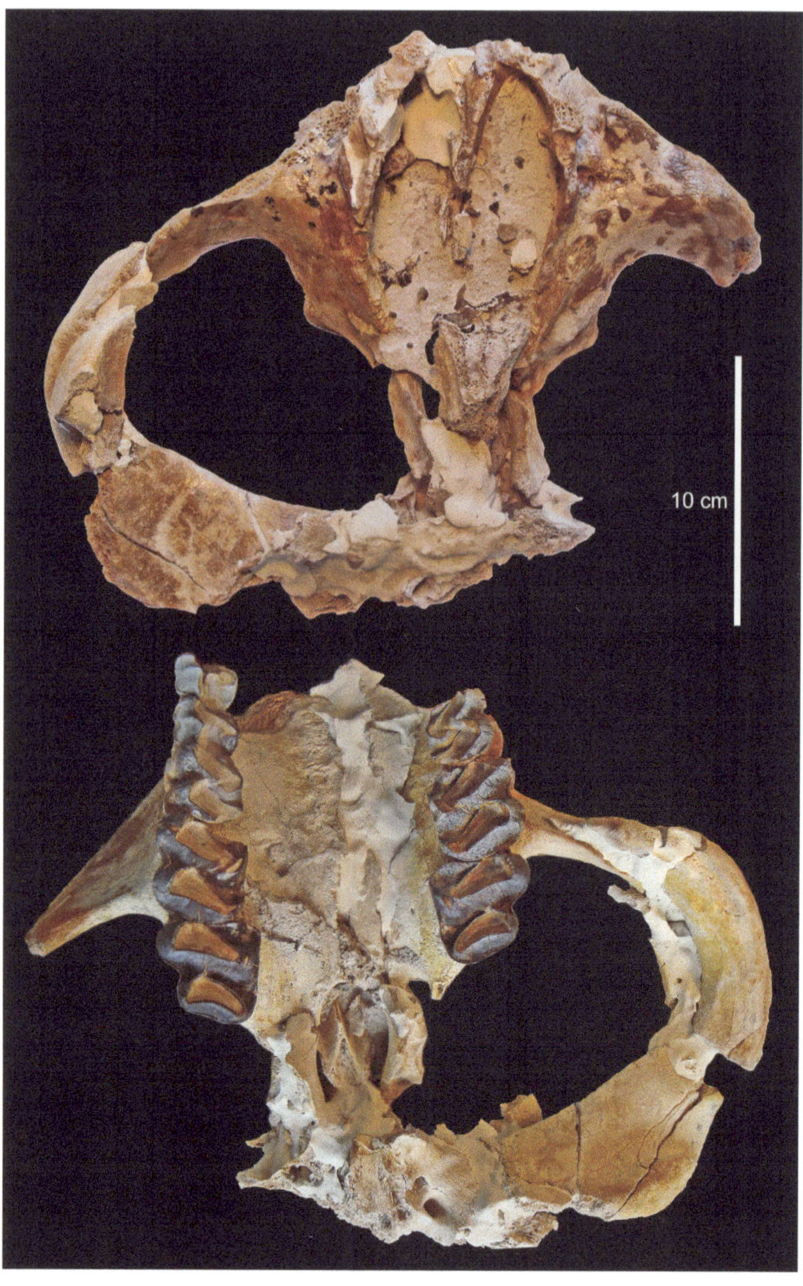

Plate 12. Skull of the embrithopod *Namatherium blackcrowense* (dorsal, *upper*, and ventral, *lower*, views) from the middle Eocene of Namibia (ca. 43 Ma). *Namatherium* was a large endemic African ungulate, the size of a large living tapir. It is a close ancestral relative of *Arsinoitherium* (family Arsinoitheriidae). From Pickford et al. (2008); photo courtesy M. Pickford.

Plate 13. Upper. The oldest anthropoid primate known in Africa is represented by a tiny upper molar (*inset*) of the species *Amamria tunisiensis*, discovered by L. Marivaux at the Djebel el Kébar fossil site in Tunisia. **Lower.** The famous toothed whale (archaeocetes) world heritage fossil sites of Wadi Al-Hitan (Fayum, Egypt). Excavation shown of the basilosaurid skeleton of *Basilosaurus isis* **(see fig. 4.21)** from the late Eocene (38–34 Ma) during the 2005 paleontological field mission of P. D. Gingerich's team. Tooth redrawn by C. Letenneur from Marivaux et al. ("Morphological Intermediate," 2014, fig. 3); photo of the Djebel el Kébar site courtesy of L. Marivaux (**upper**). Photo courtesy of P. D. Gingerich (**lower**).

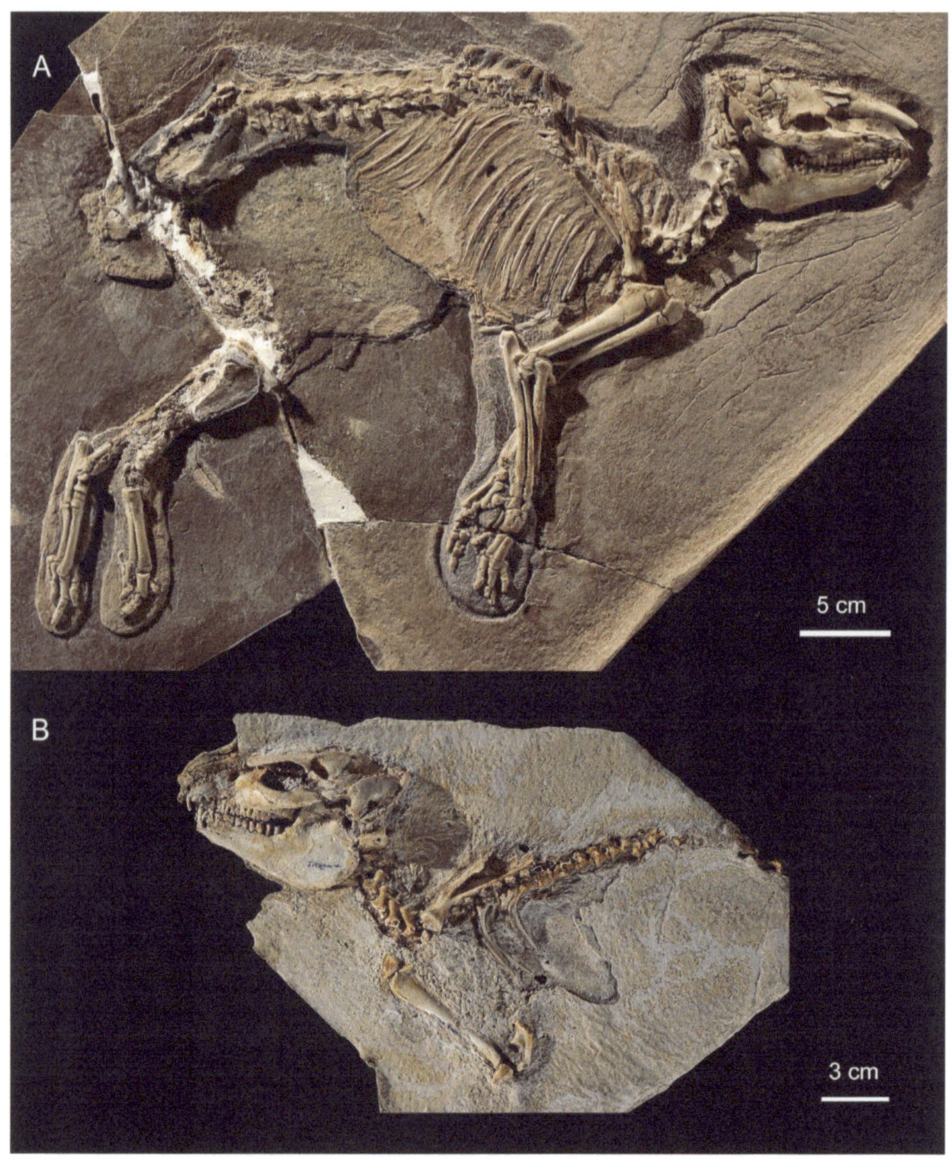

Plate 14. The early Oligocene (33 Ma) hyracoid *Saghatherium antiquum* (family Saghatheriidae) from Jebel al Hasawnah, Libya. Skeleton of an adult (**A**) and a very young individual (**B**) that still has its milk teeth. See reconstruction in **figure 4.23**. Jebel al Hasawnah is one of the very few Paleogene localities known in Africa to have yielded articulated mammal skeletons. The now-extinct *Saghatherium* was a small hyrax whose size and general appearance were quite similar to the living dassies, except that it was unguligrade and better adapted to running. From Gheerbrant et al. (2007); photos courtesy of P. Loubry.

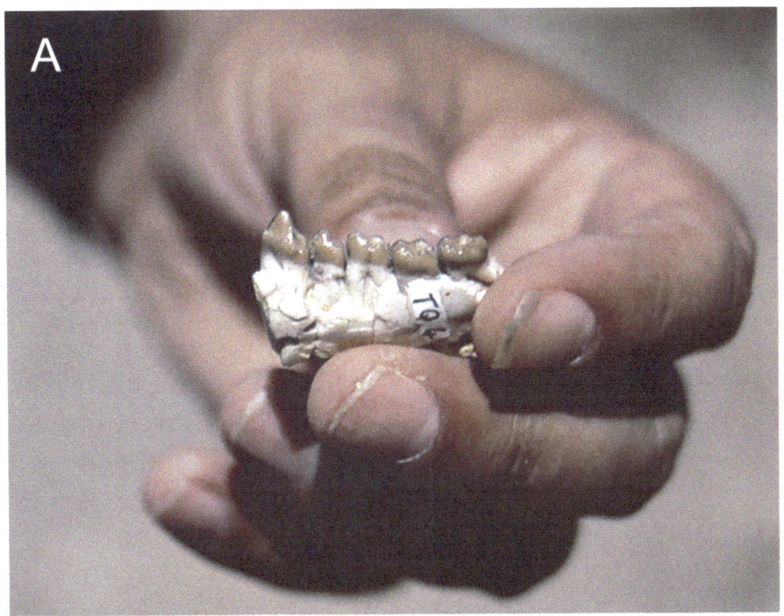

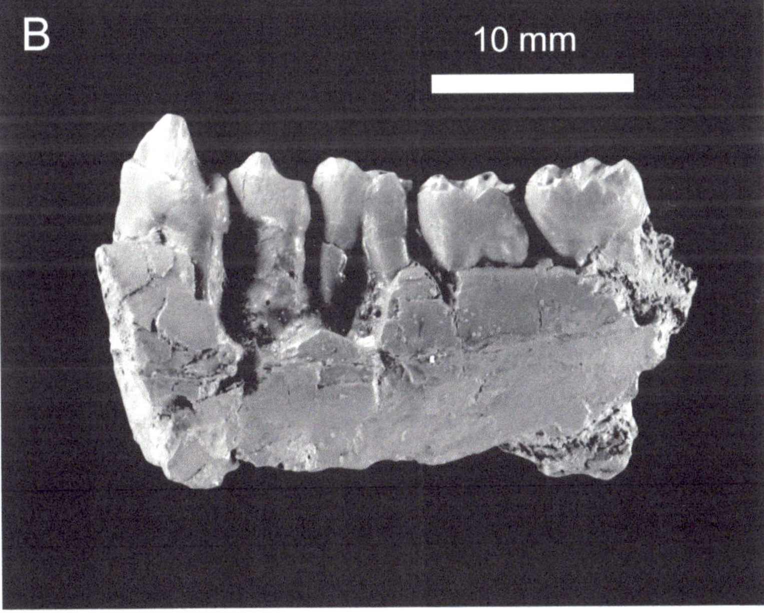

Plate 15. The stem catarrhine primate "*Moeripithecus markgrafi*" is a likely new propliopithecid species from the early Oligocene of Taqah, Oman (34 Ma). **A.** Photograph of the lower jaw as discovered in the field, in 1988, during the first excavation of the Taqah fossil site. **B.** Cast of the same specimen showing details of the premolar and molar teeth (view of the internal side). Photo courtesy of S. Sen (**A**). Photo by E. Gheerbrant (**B**).

Plate 16. The skull (view of the face) of the catarrhine primate *Saadanius*, discovered in late Oligocene beds of Arabia (29–28 Ma). *Saadanius* is the closest stem relative to the Old World monkeys and great apes. From Zalmout et al. (2010, fig. 2); photo courtesy of I. Zalmout.

CHAPTER 4

Mammal Succession and Evolution in Island Africa

Paleocene—the beginnings of the placentals

The Paleocene is the period of the initial explosive radiation of placental mammals, on almost all continents, that led to the birth of the different extinct and extant orders. Various lineages evolved rapidly, with greater or lesser success. Paleocene mammals are best documented in the Laurasian continents, particularly in North America and Asia, where the earliest fossils of rodents, primates, carnivores, and various basal placental lineages have been found. All these mammals belong to the major placental branch of the Boreoeutheria—that is, the placentals native to the Boreal regions. The most abundant Paleocene mammals are early (or even earliest) forebears of various extinct and extant placental lineages, including some major living groups. For example, there are "condylarths," which include early herbivores that either belong to extinct lineages or are ancestral to modern ungulates (Euungulata), stem insectivores that are basal to both the shrews and hedgehogs (Eulipotyphla) and a variety of extinct insectivorous lineages of uncertain relationships (Pantolestidae, Leptictidae, Palaeoryctidae), the carnivoramorphan stem groups to modern carnivores, and plesiadapiforms, an extinct lateral branch of the extant primates. Stem marsupials are also known during the Paleocene. In South America, Paleocene mammals include marsupials, "condylarths," pantodonts (which are an archaic group of herbivorous placentals), the forebears of the extinct South American endemic ungulates (litopterns, notoungulates), and a relative of the Australian platypus (monotremes). In other Gondwanan continents, such as India, Antarctica, and Australia, Paleocene mammals are completely unknown. Madagascar is even a terra incognita for the entire Paleogene period.

In Africa, our knowledge of the Paleocene fauna is poor, very patchy, and quite recent. It is based on two major Moroccan sites discovered by French paleontological teams: the Ouled Abdoun Phosphate Basin (Gheerbrant et al., 1996, 1998) and the Ouarzazate Basin (Adrar Mgorn 1 site; Cappetta et al., 1978). Mammals from these sites have been found in levels dated to the middle and late Paleocene, from 61.5–60 to 57 Ma. No earlier Cenozoic mammals are known in Africa. More broadly, no marsupial or placental—that is, no therian mammals—are known before the middle Paleocene (61.5 Ma) in Africa. About 15 mammal species belonging to 10 orders have been identified in the Paleocene levels of the Ouled Abdoun and Ouarzazate Basins. All are eutherians—that is, stem and modern (crown) placentals. There are no marsupials, nor are there archaic nontherian groups such as multituberculates. The mammals from these Moroccan sites are thus the oldest-known African placentals. The faunas of the Ouarzazate Basin and Ouled Abdoun Phosphate Basin are very different and, fortunately, complementary in their compositions.

Mammals from the Ouarzazate Basin, such as the Adrar Mgorn 1 locality **(plate 8)**, are documented only by isolated teeth of several species. Three hundred small, isolated teeth were collected after a long and tedious lab process using acid preparation of large quantities of hard fossiliferous limestone, several tons of which were sampled in the field. The teeth almost all belong to "micromammals" with millimetric teeth. There are diminutive archaic insectivore-like mammals belonging to at least 10 species studied in the PhD thesis of the author (Gheerbrant, 1992, 1994, 1995). They represent basal placental lineages (eutherians) and stem groups of various modern insectivorous and carnivorous placentals. *Cimolestes cuspulus* (family Cimolestidae) is likely close to the root of the hyaenodont lineage of extinct predators, which includes the only carnivorous mammals that evolved in Island Africa. *Tinerhodon disputatum* is more specialized than *Cimolestes cuspulus*. It is the basal-most known hyaenodont. For some authors, it is related to the origin of the European proviverrines. *Tinerhodon* has the sharp sectorial lower molars of the hyaenodonts with a specialized large bladelike tubercle at their front. Unfortunately, this species is not yet well known, and we know only its lower teeth.

Todralestes variabilis (family Todralestidae) is by far the most abundant species in the Ouarzazate Basin sites, representing 70% of recovered mammal fossils (45% in terms of number of individuals). It may have lived in the local mangroves of the Moroccan coastal regions, which are represented by fossil wood and fruits found in the region. *Todralestes* was described as a stem placental (eutherian) possibly related to the extinct Pantolesta group (Gheerbrant, 1994). Pantolestan mammals, like palaeoryctids, are actually of uncertain relationships, being either stem placentals or stem eulipotyphlans. They are well known in Europe and North America during the Paleocene. They evolved in a semiaquatic mode of life and are specialized for a diet of hard prey, such as crabs, shrimps, and shellfish. However, there is no known aquatic specialization in *Todralestes*. Interestingly, the more recent study of Seiffert (2010) concluded that *Todralestes* is not a stem eutherian but a stem tenrecoidean (= afrosoricidan) ancestral to the golden moles, tenrecs, and potamogales (see also Heritage et al., 2021). However, *Todralestes* does not display the specialized morphological traits of the tenrecoideans, such as the puncturing zalambdodont molars designed to pierce insect shells. In fact, and intriguingly, in *Todralestes* it is the enlarged and sharp last upper premolars that are specialized to pierce insect shells **(fig. 4.1)**.

Another insectivore-like placental is represented by a tiny species, *Palaeoryctes minimus*, of the Paleocene family Palaeoryctidae. This family, best known from Laurasia, has specialized puncturing zalambdodont molars resembling those of golden moles and tenrecs, although in a less advanced stage. As its species name indicates, *Palaeoryctes minimus* from the Ouarzazate Basin is very small. Its teeth are less than a millimeter long. From the size of the teeth, its body mass was estimated to be two grams, making it one of the smallest-known mammals.

An extinct enigmatic family of basal insectivorous eutherian mammals, the Adapisoriculidae is present in the Ouarzazate Basin. It is represented by the species *Afrodon chleuhi*. Adapisoriculids were small arboreal

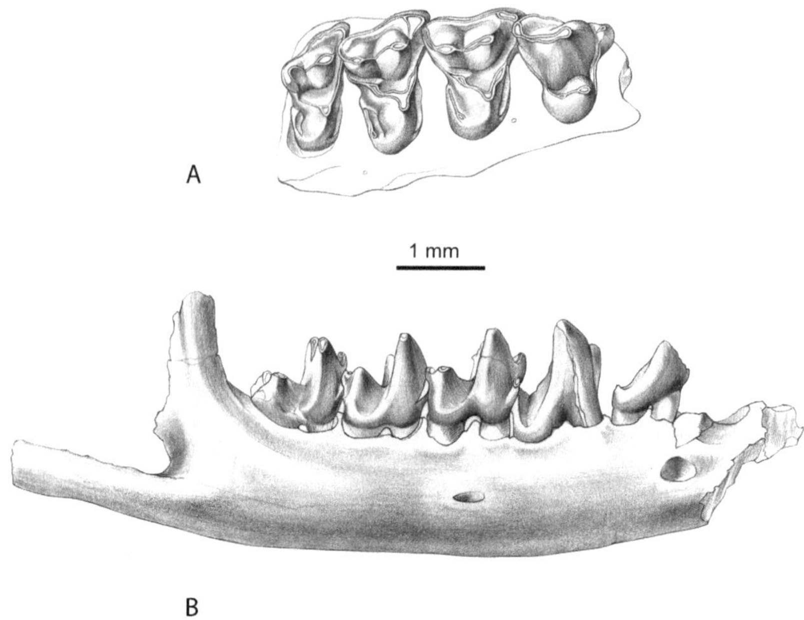

Figure 4.1. (**A**) Upper jaw (bottom view), and (**B**) lower jaw (side view) of the stem insectivorous placental mammal *Todralestes variabilis* from the late Paleocene of the Adrar Mgorn site, Ouarzazate Basin, Morocco, 57 Ma. This is the best-known mammal from the site (150 specimens identified, 45% of the mammal species frequency)—all other mammal species being known from a few isolated fragmentary teeth recovered by the acid attack of several tons of the hard limestone that composes the fossiliferous layer. Redrawn by C. Letenneur from Gheerbrant (1994, figs. 1 and 10).

mammals that climbed on branches with their claws like squirrels. In Africa adapisoriculids are known until the early and likely middle Eocene (Black Crow site). The family is also known at the end of the Cretaceous in India and during the Paleocene in Europe. Its paleogeographic distribution indicates ancient intercontinental faunal interchanges with Island Africa, the nature of which remains poorly understood: We don't know whether they arrived on rafts of floating plant debris or by terrestrial routes via more or less continuous land bridges. In fact, the very origin of the Adapisoriculidae family is mysterious. It dates to the Cretaceous, and it may be sought in a Gondwanan continent, such as India or Africa (Goswami et al., 2011). A tooth from the Adrar Mgorn 1 site (**fig. 4.2**)—unique but interesting—shows morphological relationships to afrotherian insectivores of the extant African tenrecoid group that includes the tenrecs,

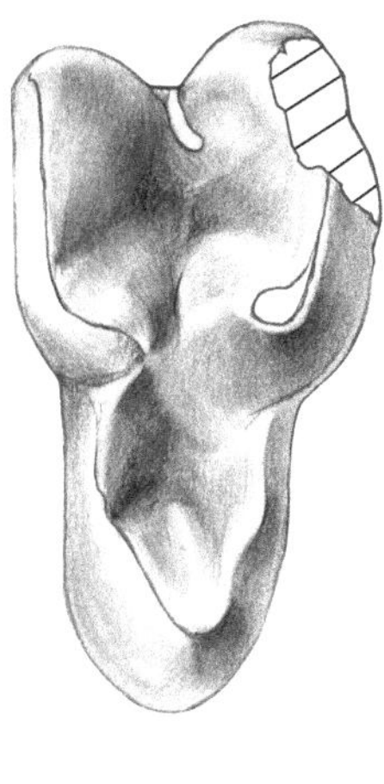

Figure 4.2. Mammal upper molar (specimen THR195, viewed from below) from the Adrar Mgorn locality, Ouarzazate Basin, Morocco, 57 Ma. This tooth is the only one known from a species (still unnamed) possibly belonging to tenrecoid afrotherians. It might represent the earliest-known afrotherian insectivores, which include the extant tenrecs, golden moles, and potamogales. Redrawn by C. Letenneur from Gheerbrant (1995, fig. 22).

otter shrews, and golden moles. Like them, this species has transversely enlarged upper molars with a wide stylar shelf and a low and elongated inner tubercle (protocone). This tooth could be the oldest-known fossil milestone of the tenrecoideans.

The most important discovery in the Paleocene sites of the Ouarzazate Basin, for which the locality is famous, is that of the species *Altiatlasius koulchii*, described in 1990 by Bernard Sigé (University of Montpellier) and colleagues **(fig. 4.3)**. This small species, weighing 120 grams, is known by only a dozen teeth with typical bunodont crushing morphology in which the tubercles and crests are swollen and low. *Altiatlasius* is the oldest-known true primate (Euprimates). Its relationship to other primates remains uncertain because of the limited nature of the available fossils. *Altiatlasius* most likely arrived in Africa following an earlier immigration from Asia during the Paleocene.

The Paleocene mammals discovered by the author in the phosphate deposits of the Ouled Abdoun Basin (Morocco) are somewhat older, of

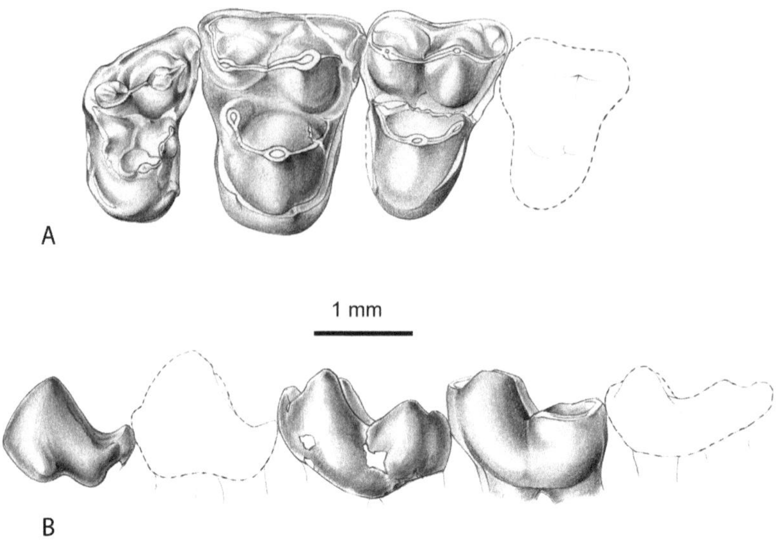

Figure 4.3. One of the earliest-known true primates (euprimates): *Altiatlasius koulchii* from the late Paleocene of Adrar Mgorn site, Morocco, 57 Ma. Upper (**A**) and lower (**B**) cheek teeth rows of *Altiatlasius* (bottom and side views, respectively) reconstructed from isolated teeth found in the site. *Altiatlasius* was a diminutive stem primate. Redrawn by C. Letenneur from Sigé et al. (1990, plate 4).

Selandian and Thanetian age (about 60 to 57 Ma), and they are quite different from those of the Ouarzazate Basin. They mainly come from quarries in the eastern part of the Ouled Abdoun Basin (Daoui, Meraa El Arech, and Sidi Chennane areas). Mammal fossils are very rare at these sites, which are instead famous for their rich and abundant marine fauna (sharks, bony fishes, turtles, crocodiles). The Ouled Abdoun sites are nevertheless very important because the fossils are more complete and better preserved than those of the Ouarzazate Basin: They consist of jaws and even some skulls, which are very informative. The fossils belong to medium-sized species (from 4 to 12 kg), larger than those from the Ouarzazate Basin, and they belong to different groups. This difference in the faunal composition of the sites of the two Moroccan basins is due to different sedimentary environments, although they were both infilled by coastal marine sediments. The dominant mammal group found in the Ouled Abdoun Basin is composed of ungulate-like afrotherians (i.e., paenungulates). There are at least four species belonging to two distinct paenungulate groups: modern (extant) and ancestral paenungulates. Modern paenungulates include the living dassies, elephants, and sea cows, as well as their early relatives. Mammals from the Ouled Abdoun Basin are the oldest paenungulates and afrotherians known in the world.

A remarkable species is *Eritherium azzouzorum* (Gheerbrant, 2009) **(fig. 4.4)**. It is the oldest-known species of the elephant order, the proboscideans. It has an archaic morphology and a small size (body mass of about 4–5 kg; shoulder height of about 20 cm)—quite distinct from modern elephants! Yet *Eritherium* shares morphological details unique to the elephant group. In particular in the structure of the orbit of the skull (with the maxillary bone extended under the orbit) and the teeth (bunodont-lophodont molar, enlarged first incisor). Its primitive morphology documents a very early evolutionary stage of emergence of the proboscideans. In fact, it is a practically unprecedented case. Few fossils of ancient species are known that record the emergence of modern placental orders at such an early phase of their evolution. And for the order of elephants—which are among the most specialized and spectacular mammals—this is quite remarkable. As such, *Eritherium* is one of the most demonstrative pieces of fossil evidence supporting Charles Darwin's theory of "descent with modification." *Eritherium* provides key information about the evolution of African ungulates. It indicates that the three extant paenungulates orders comprising the dassies, elephants, and sea cows were already well differentiated in the middle Paleocene, by 60 Ma, and that their evolutionary history dates to at least the beginning of the Cenozoic. It shows that the diversification of African ungulates (paenungulates) took place rapidly at the dawn of the Age of Mammals (Gheerbrant, 2009).

The first African "condylarths," which unveil a previously unknown chapter in the evolution of African ungulates, were also discovered in the Ouled Abdoun Basin. These African "condylarths" are related to the ancestry of all extant African orders of ungulates: They are stem paenungulates. They were discovered in 2001 with the two species *Abdounodus hamdii* **(fig. 4.5)** and *Ocepeia daouiensis* (Gheerbrant et al., 2001)

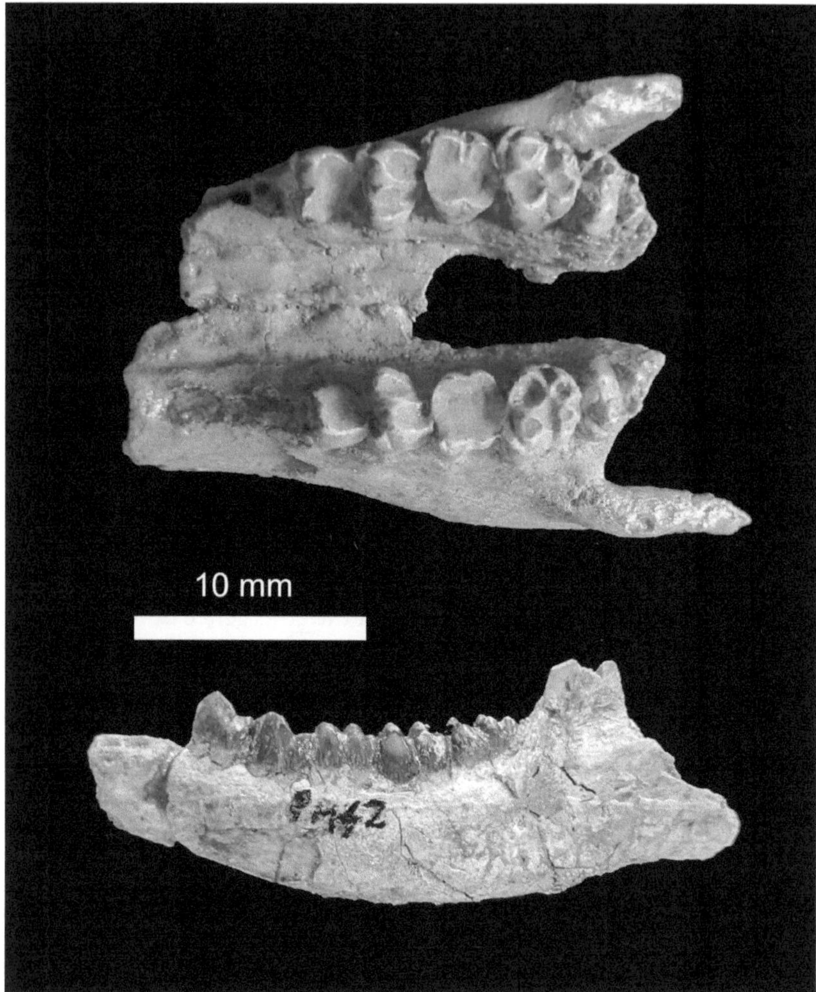

Figure 4.4. *Eritherium azzouzorum*, the oldest-known elephant relative (proboscideans). Upper (*top*) and lower (*bottom*) jaws found in a middle Paleocene bed (60 Ma) of the Ouled Abdoun phosphate quarries, Morocco. *Eritherium* is an archaic (plesiomorphic) proboscidean, close in morphology to the hyracoids, and as small as extant dassies. However, it shares some characters that are unique to the proboscideans. It is the witness of the origin and the rapid basal diversification of the proboscideans and other paenungulate orders at the dawn of the Cenozoic. Photo by E. Gheerbrant.

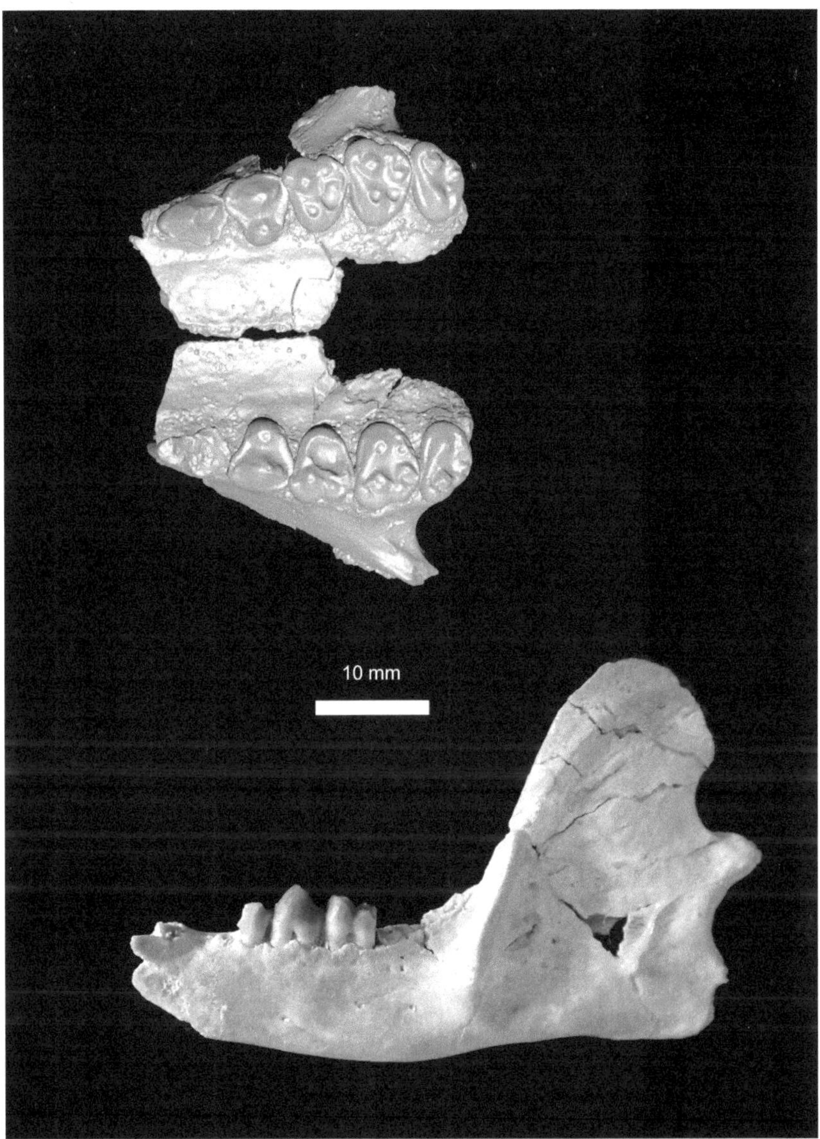

Figure 4.5. Maxilla (*top*) and mandible (*bottom*) of the African "condylarth," or stem paenungulate, *Abdounodus hamdii* from the middle Paleocene of the Ouled Abdoun Phosphate Basin, Morocco, circa 60 Ma. Photo by E. Gheerbrant.

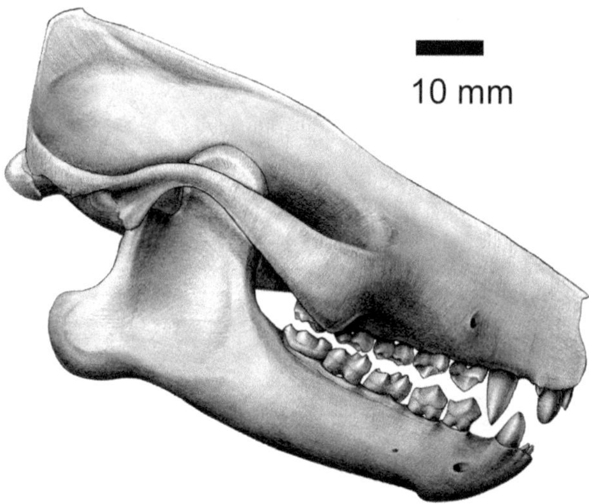

Figure 4.6. The reconstruction of the head and skull (side view) of *Ocepeia daouiensis* based on the 3D digital modeling of CT scans of the skull and the mandible shown in plate 9. Drawing by C. Letenneur from Gheerbrant et al. (2014, figs. 12 and 13).

(fig. 4.6; plate 9). At that time, they were poorly known by three fragments of lower jaws. Since then, the author has reported additional and better-preserved fossils, including a skull of *Ocepeia daouiensis* that is the only known skull of a mammal in the Paleocene of Africa (Gheerbrant et al., 2014). *Ocepeia daouiensis* and *Abdounodus hamdii* are small herbivorous mammals with bunodont (swollen tubercles) crushing teeth. Another stem paenungulate, represented by the species *Hadrogeneios phosphaticus*, was recently described (Gheerbrant, 2023). It is the basal-most known stem group of paenungulates. It displays a strangely specialized mandibular symphysis that is enlarged and raised, and has a curious shape like pliers.

Hadrogeneios, Ocepeia, and *Abdounodus* provide the only known fossil evidence of the early evolution and diversification of African ungulate mammals before the origin of extant paenungulate orders (dassies, elephants, sea cows). *Ocepeia* and *Abdounodus*, in particular, illustrate previously unknown intermediate morphological stages in the evolution of molars with four tubercles (quadritubercular molars), which are found in all modern paenungulates. Molars with four tubercles evolved in modern paenungulate orders as the first morphological step toward bilophodont molars. On the bilophodont teeth, the tubercles are connected in pairs by a transverse crest with a cutting function. This specialized morphology is adapted to a folivorous (leaf-eating) diet, widely found in ungulates. The intermediate morphology of *Ocepeia daouiensis* and *Abdounodus hamdii* shows that the basic morphological plan of the quadritubercular molar is not homologous in paenungulates and Laurasian euungulates such as perissodactyls (horses, rhinoceros): It is constructed differently. Thus, *Ocepeia daouiensis* and *Abdounodus hamdii* demonstrate that the quadritubercular and bilophodont molars evolved convergently in the African and Laurasian ungulates (Gheerbrant et al., 2016) as an adaptation to a similar folivorous diet and herbivorous niche. The Selandian phosphate levels (60 Ma) of the Ouled Abdoun Basin also yielded a small carnivorous mammal, the hyaenodont *Lahimia selloumi* **(fig. 4.7)**. Paradoxically, this species is more specialized than the younger hyaenodont species *Tinerhodon disputatum* from the Thanetian, circa 57 Ma, of the Ouarzazate Basin. *Lahimia* is not only the oldest-known

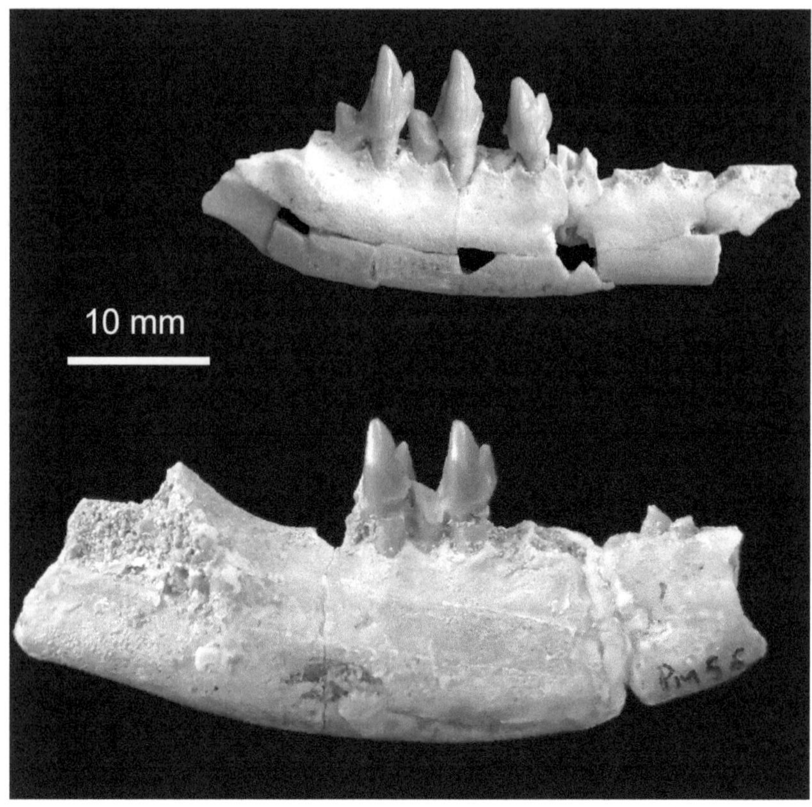

Figure 4.7. *Lahimia selloumi* (Koholiinae) is the earliest carnivorous mammal known in Africa. It belongs to the extinct hyaenodont order that evolved and diversified in Island Africa. It is known by several partial mandibles found in the middle Paleocene marine fossiliferous phosphate beds of the Ouled Abdoun Basin, Morocco, circa 60 Ma. Hyaenodonts are extinct carnivorous mammals unrelated to the Laurasian order of the Carnivora. Photo by E. Gheerbrant

hyaenodont; it is also a representative of a specialized African lineage, the Koholiinae. *Lahimia* bears witness to an already ancient evolution of hyaenodonts in Africa, since the early Paleocene.

What do Paleocene fossils tell us about the origin and early history of the African mammals? First, note that the earliest-known African Cenozoic mammal faunas are characterized by a mixture of cosmopolitan ancestral groups (stem placentals) and endemic modern (extant) groups. The "primitive" African fauna is represented by the small, generalized insectivore-like mammals found in the Ouarzazate Basin sites (e.g., Adrar Mgorn 1).

These are mammals of preplacental grade (eutherians: Cimolesta, Adapisoriculidae) and stem groups of various placental lineages, some extinct, such as the hyaenodonts, and other extant, such as the afrotherians and the primates. This assemblage corresponds to an early ancestral stock of mammals of Laurasian (mostly Eurasian) origin that dispersed to Africa at the end of the Cretaceous or at the beginning of the Cenozoic. It includes the first-known lineages of stem and crown (modern) placentals to arrive in Island Africa. The most derived mammals known in the Paleocene of Africa are forebears of endemic lineages such as afrotherians. Some are ungulate-like afrotherians (i.e., paenungulates); they include both stem paenungulates (*Hadrogeneios*, *Ocepeia*, *Abdounodus*) and derived paenungulates (e.g., the proboscidean *Eritherium*). Others are probable stem groups of afrotherian insectivores, such as the tenrecoideans, which include tenrecs, otter shrews, and golden moles. These mammals are the witnesses of an already-advanced autochthonous evolution—of the antiquity of placentals in Island Africa. This is in line with the placental phylogeny, which shows that Afrotheria is one of the first branches of the placentals. That said, a big question remains: How old are the earliest placentals of Africa? In the absence of fossils earlier than the middle Paleocene, we do not know.

The modern aspect of the ancient African fauna is quite original. Known Paleocene representatives of modern placentals (i.e., extant placental orders) remain rare elsewhere, including in North America and Asia, where fossils are the most abundant (and much more abundant than in Africa). They generally appear in the fossil record only from the late or latest Paleocene (57–56 Ma). They include the insectivore mammals such as shrews and hedgehogs (eulipotyphlans), the rodents and the carnivores found in North America, and the primates and rodents found in Asia.

Eocene—the rise of emblematic African groups

Early Eocene, 56–48 Ma

The early Eocene is a poorly documented period in Africa despite important recent discoveries. Three main sites have been discovered in Algeria

(El Kohol) and Morocco (Ouled Abdoun and Ouarzazate Basins). These sites yielded 12 species of seven orders of placental mammals. Some of them represent old native lineages known in the earlier Paleocene period: These are archaic stem elephants (phosphatheriid proboscideans), hyaenodonts (koholiines), and insectivore-like eutherians representing pre-placental lineages. Other groups found in the early Eocene sites make their first appearance in Africa: They are the first-known hyracoids (*Seggeurius*) and embrithopods (*Stylolophus*), as well as some stem bats. Hyracoids and embrithopods, together with elephants and sea cows, all belong to the African ungulate group of paenungulates. Their absence in Paleocene beds almost certainly corresponds to gaps in the fossil record. The discovery of the middle Paleocene proboscidean *Eritherium* indicates that the hyracoids and embrithopods, which are the most basal lineages to the modern (extant) paenungulates, originated at least in the beginning of the Paleocene.

As in the Paleocene, most known early Eocene mammals are small. For example, the stem proboscidean *Phosphatherium escuilliei* (Gheerbrant et al., 1996) and the embrithopod *Stylolophus minor* (Gheerbrant et al., 2018) are notoriously small for their group, with sizes close to a medium-sized dog (estimated body mass 15 kg) and goat (ca. 30 kg), respectively. According to the known fossil record, as incomplete it is, there are some extinctions at the Paleocene–Eocene boundary (56 Ma), including that of the African "condylarths" (i.e., the stem paenungulates) such as the ocepeiids. However, apart from these few extinctions, mammal evolution in Africa across the Paleocene–Eocene transition is characterized by the presence of ancient lineages and a rapid evolution of some of them, especially the extant orders of African ungulates (paenungulates). The most striking example is that of the proboscideans, which evolved rapidly with large animals such as *Daouitherium* (body mass about 200 kg) and *Numidotherium* (body mass about 250 kg; shoulder height of 1 m). This evolution has been related by Yans and colleagues (2014) to successive early Eocene events of intense climatic warming and, in particular, to the so-called Early Eocene Climatic Optimum (EECO) middle Ypresian event, which is dated 53–51 Ma, the age of *Daouitherium*. The relative continuity of Paleocene–Eocene evolution in Africa is in contrast with evolution in the Laurasian domains (northern continents) where the Paleocene–Eocene boundary (56 Ma) was a period of dramatic faunal change, with numerous disappearances and major appearances linked to large intercontinental dispersals (geodispersals) and colonizations of mammals. The Paleocene–Eocene faunal event in Laurasia is linked to a major climatic warming event called the Paleocene–Eocene Thermal Maximum (PETM), which seems to have had more impact in Laurasia than in Island Africa.

A few Paleocene lineages of insectivore-like stem or early placentals survived in the Ypresian of the Ouarzazate Basin. These include the three families palaeoryctids, adapisoriculids, and todralestids, all being tiny shrew-like mammals. There are also small carnivorous mammals of the extinct hyaenodont order, in particular the species *Boualitomus marocanensis* and *Koholia atlasense*, which represent a specialized and ancient African predator lineage, the Koholiinae.

Seggeurius amourensis from the El Kohol site **(fig. 4.8)** and another species of *Seggeurius* from the Ouled Abdoun Basin phosphates are the oldest-known hyracoids (dassies order). They are small and plesiomorphic, like other African ungulates (paenungulates) of the same age. *Seggeurius amourensis* is known by a broken skull that displays little specialized morphology (Benoit et al., 2016). The molars are bunodont, with low and inflated tubercles and crown, and the premolars are simple. The poorly specialized morphology of *Seggeurius amourensis* places it at the root of the hyracoids. However, the relationship of hyracoids, the first branch of the paenungulates in the phylogenetic tree, indicates a much older origin of the group, at least from the beginning of the Paleocene. It implies a lack of fossils over a time span of at least 10 million years in their long evolutionary history.

Phosphatherium **(fig. 4.9; plate 10)** from the early Ypresian (56–55 Ma) of the Ouled Abdoun Basin (Morocco) is a small proboscidean (shoulder height of 30 cm; body mass of 15 kg) known from several fossil skulls (Gheerbrant et al., 2005). It is the forebear of the stem lophodont proboscideans bearing cutting teeth, which underwent a limited diversification in the middle and late Eocene, the first one in the evolution of proboscideans. The lophodont teeth evolved as the oldest dental specialization in the elephant order. They are characterized by the

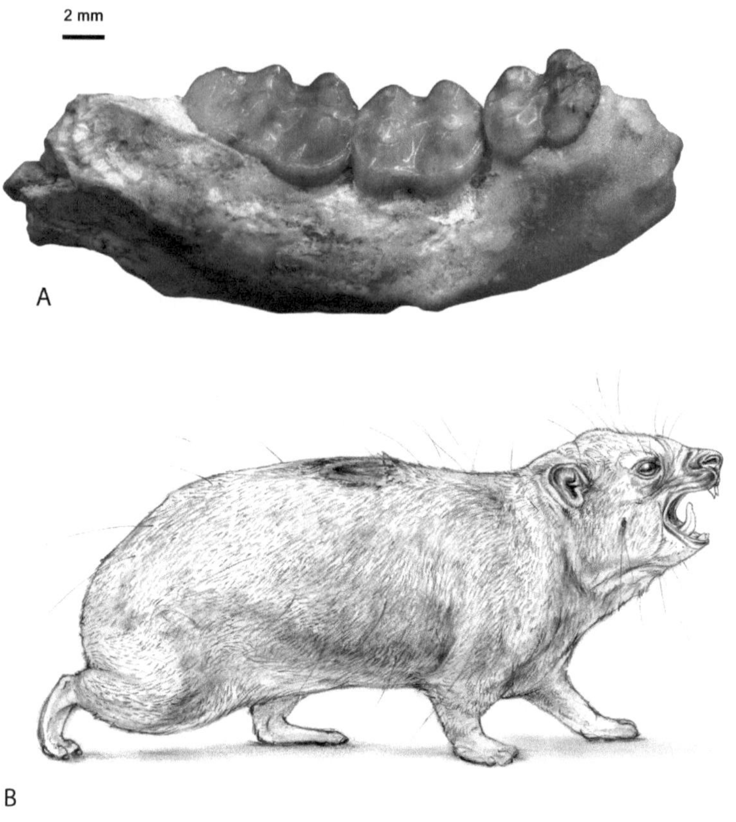

Figure 4.8. A. Partial lower jaw with teeth (type specimen) of the earliest-known hyrax, *Seggeurius amourensis*, discovered in the early Eocene beds of El Kohol site, Algeria, 51 Ma. **B.** The living rock hyrax, *Procavia capensis* (body length 30–50 cm; body mass 2–5 kg). From Court and Mahboubi (1993, fig. 4); photo courtesy of R. Tabuce (**A**). Drawing by C. Letenneur (**B**).

presence of long transverse cutting crests on the teeth that, during mastication, function between opposed teeth, like a pair of scissors. These teeth correspond to a specialized folivorous (leaf-eating) diet. The lophodont tooth is a key adaptive innovation that led to the first evolutionary success of proboscideans during the Eocene, resulting in elephant-sized mammals such as *Barytherium*. Early elephants with lophodont teeth evolved into several successive lineages that root at the base of all modern groups. They do not form a single natural group (i.e., they are not a clade). *Daouitherium rebouli* from the middle Ypresian (53 Ma) of the Ouled Abdoun Basin (Gheerbrant et al., 2002) and *Numidotherium*

Figure 4.9. The early elephant relative *Phosphatherium escuilliei* (56 Ma) from the Ouled Abdoun Phosphate Basin (Morocco). **A.** Skull reconstruction. **B.** Life restoration. 3D plaster model reconstructed by D. Visset from casts of fossils; photo by E. Gheerbrant (**A**). Drawing by C. Letenneur (**B**).

koholense from the late Ypresian (51 Ma) of El Kohol (Mahboubi et al., 1984) belong to this initial bushing of proboscideans with lophodont teeth, and they were the first large mammals to evolve in Africa. *Daouitherium* is only known by its lower jaw (**fig. 4.10**). *Numidotherium koholense* is one of the best-known Paleogene mammals from Africa. Abundant fossils have been found, including a skull (**plate 11**) and

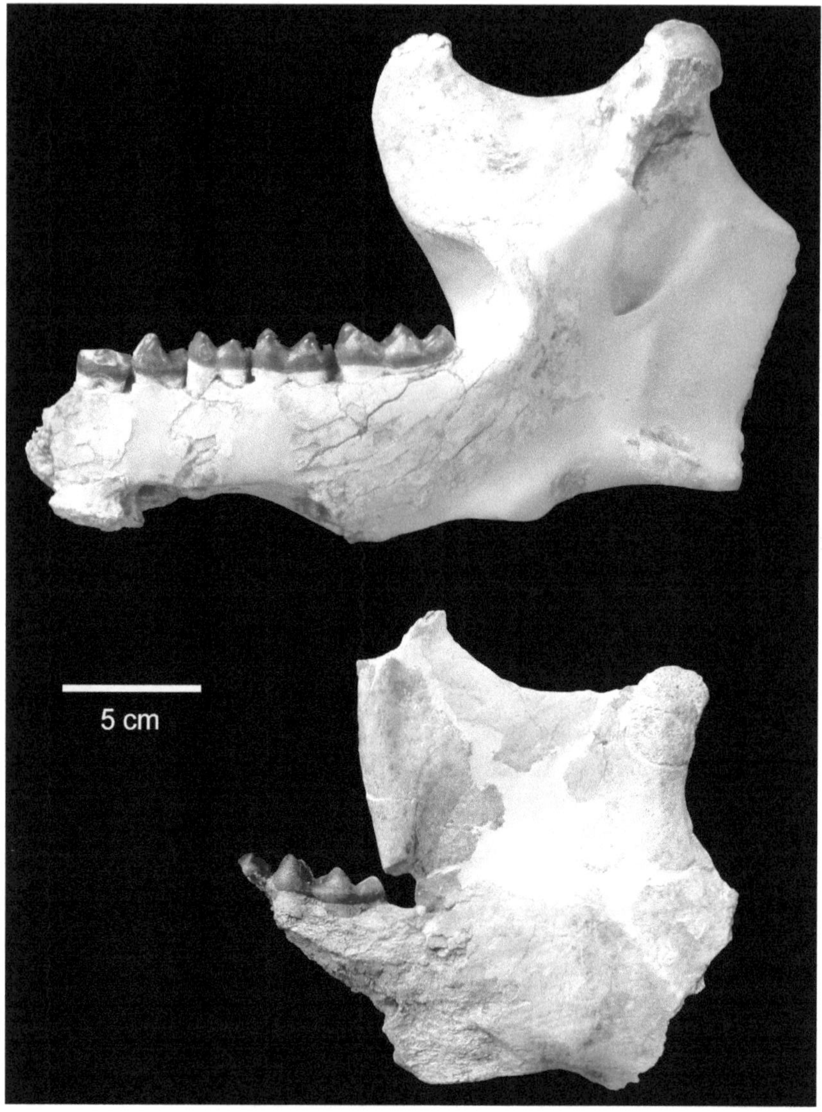

Figure 4.10. The earliest-known large proboscidean, *Daouitherium rebouli*, from the early Eocene beds (53 Ma) of the Ouled Abdoun Basin phosphate quarries, Morocco. Views of the lower jaw with cheek teeth. With a body mass of about 200 kilograms, *Daouitherium* was the size of a tapir. Photos by E. Gheerbrant

many other bones of the skeleton (Mahboubi et al., 1986; Noubhani et al., 2008). The fossils of *Numidotherium koholense* record a local population of numerous male and female individuals of all ages, from young to old. The skull shows notable variability in size and morphology, indicating that *Numidotherium koholense* was a sexually dimorphic species. The sexual dimorphism seen in *Numidotherium koholense* suggests it was gregarious, living in social groups with sex-related behavioral roles (i.e., polygynous species). The morphology of the limb bones of *Numidotherium* indicates that it was a stem plantigrade proboscidean with laterally flexed limbs (abducted limb) (Court, 1994), unlike modern elephants, which have raised columnar limbs (graviportal skeleton) and are unguligrade.

The paenungulates of the extinct embrithopod order appear in the early Eocene with the genus *Stylolophus*, recently discovered in the Ouled Abdoun Basin in Morocco (Gheerbrant et al., 2018, 2021). *Stylolophus* is represented by two successive species of increasing size, *S. minor* (body mass about 30 kg, or about the size of a small goat) and *S. major* (body mass about 90 kg, or about the size of a wild boar), found in two phosphate levels respectively dated from the early (56 Ma) and middle (53 Ma) Ypresian. *S. minor* is the better-known species; it is represented by several broken bones of the skull and jaws **(fig. 4.11)**. The CT scan study and 3D modeling of these bones allowed the reconstruction of part of its skull, such as the jaws and snout **(fig. 4.12)**. The species *S. minor* has an enlarged first incisor and a primitive placental dental formula (i.e., three incisors, one canine, four premolars, and three molars in each half jaw).

The embrithopods are extinct large herbivores known only from the Paleogene of Africa, Romania, Croatia, and Turkey. They are close relatives of sea cows and elephants, with which they form a particular paenungulate group called the Tethytheria. The best-known and most famous embrithopod is *Arsinoitherium*, from the late Eocene and the Oligocene of Africa, which vaguely resembles a rhinoceros because of its large size and the presence of nasal horns. The embrithopods belong to a paenungulate lineage that branched off before the evolution of the first proboscideans and sirenians. *Stylolophus*, by its antiquity and its

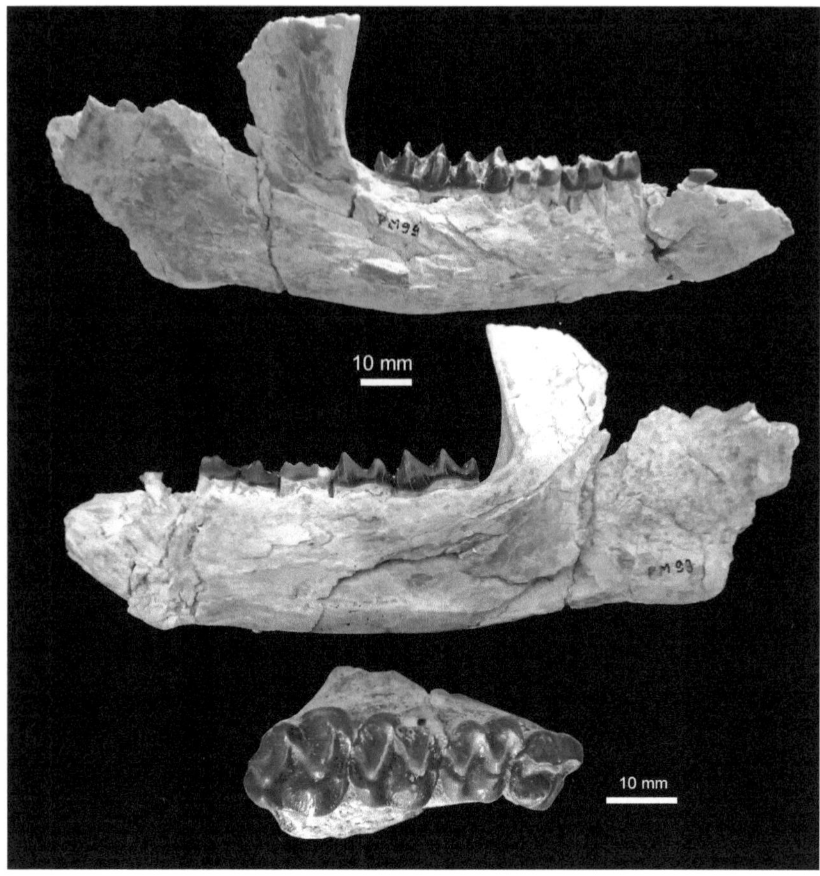

Figure 4.11. *Stylolophus minor*, the oldest and most basal known embrithopod. Lower jaw (holotype) (*top and middle*) and maxilla (*bottom*) with teeth. *Stylolophus minor* was coeval with the early proboscidean *Phosphatherium escuilliei* but larger than it, with a body size close to that of a goat (body mass of 30 kg). It was discovered in the early Eocene (56 Ma) beds of the phosphate quarries in the Ouled Abdoun Basin, Morocco. Photos by E. Gheerbrant

primitive morphology, demonstrates the African origin of embrithopods. The family Palaeoamasiidae, best known from the Eocene of Turkey and Romania, originated from an ancestral species close to *Stylolophus* that crossed the Tethys Sea northward. *Stylolophus* displays the specialized tooth morphology unique to embrithopods. These are the "hyperdilambdodont" upper molars, which bear W-like crests that are extended and enlarged transversely. The already highly specialized

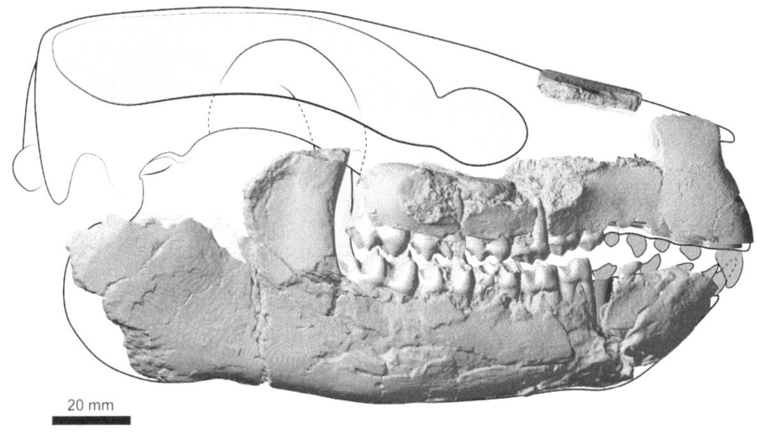

Figure 4.12. Partial reconstruction of *Stylolophus minor* (56 Ma) skull (side view) after 3D digital modeling from CT scans of fossils. From Gheerbrant et al. (2021); 3D digital image reconstructed by F. Goussard from CT scans; drawing by C. Letenneur.

dental pattern of *Stylolophus*, together with the sister group relationship of the embrithopods to both the proboscideans and sirenians, indicates that the embrithopods have an ancient evolutionary history going back at least to the Paleocene. It is intriguing to note that the first embrithopods (e.g., *Stylolophus*) were larger than the first proboscideans (e.g., *Phosphatherium*).

Three isolated small mammal teeth found in the El Kohol site (about 51 Ma) in Algeria represent the oldest-known African bat (Ravel et al., 2011). They belong to a poorly known species of stem bat ("eochiropteran") with no direct relationship to modern bat families.

Middle Eocene, 48–38 Ma

Middle Eocene mammals are well documented in Africa during the Lutetian (48–41 Ma) with about 50 species reported in some 10 fossil localities. The most important Lutetian sites are those of Chambi (Tunisia; 48 Ma), Gour Lazib (Hamada du Draa, Algeria; 48 Ma) and Black Crow (Namibia; about 43 Ma), although it should be noted that the exact age of Chambi and Gour Lazib remains uncertain—they are either latest Ypresian or earliest Lutetian. The later middle Eocene Bartonian stage (41–38 Ma) is one of the most poorly known periods concerning the paleontology of the Cenozoic African mammals. Only five Bartonian mammal species have been identified, and these are mostly marine. The single known Bartonian continental site is that of Djebel el Kébar, Tunisia, studied by Laurent Marivaux.

Compared to previous periods, the diversity of the middle Eocene mammals in Island Africa takes a leap forward. Mammal lineages present in earlier times underwent a rapid diversification during the Lutetian, which corresponds to the beginning of the great endemic African radiation. Carnivorous mammals of the extinct hyaenodont order, hyracoids, and archaic elephants (proboscideans) made their first diversification at this time. There were also several important first appearances in the Lutetian. Some are probably only pseudoappearances related to an incomplete earlier fossil record. This is the case for elephant shrews (macroscelideans), golden moles (afrotherian insectivores), sea cows (sirenians), and primates of the lemur branch (strepsirrhines). Their relationships indicate that they evolved earlier in Africa, even though fossils older than the Lutetian are not yet known. Other mammals that appear in the middle Eocene of Africa, such as toothed whales (archaeocetes) and rodents (two groups), are issued from a more or less recent (early to middle Eocene) dispersal in Island Africa. Modern bats also appear in Africa in the middle Eocene.

A single tooth, an upper molar of *Kasserinotherium* from the Chambi site, has been described as a possible African peradectid marsupial. Peradectids are stem marsupials. The tooth from Chambi would be the oldest-known relative of the marsupials from Africa. However, its identification is uncertain and needs to be confirmed by additional fossils.

The first diversification of the carnivorous hyaenodont mammals in Island Africa occurred during the Lutetian, at about 48 Ma. It is a basal radiation of the endemic African group Hyainailouroidea. It is represented by five species from Tunisia and Algeria of the genera *Parvavorodon*, *Glibzegdouia*, and *Furodon* (**fig. 4.13**) (Solé et al., 2014, 2016), which are the earliest-known hyainailouroids and their stem groups. They vary in size, morphology, and adaptations. They were specialized terrestrial hypercarnivores, and usually relatively large. *Glibzegdouia*

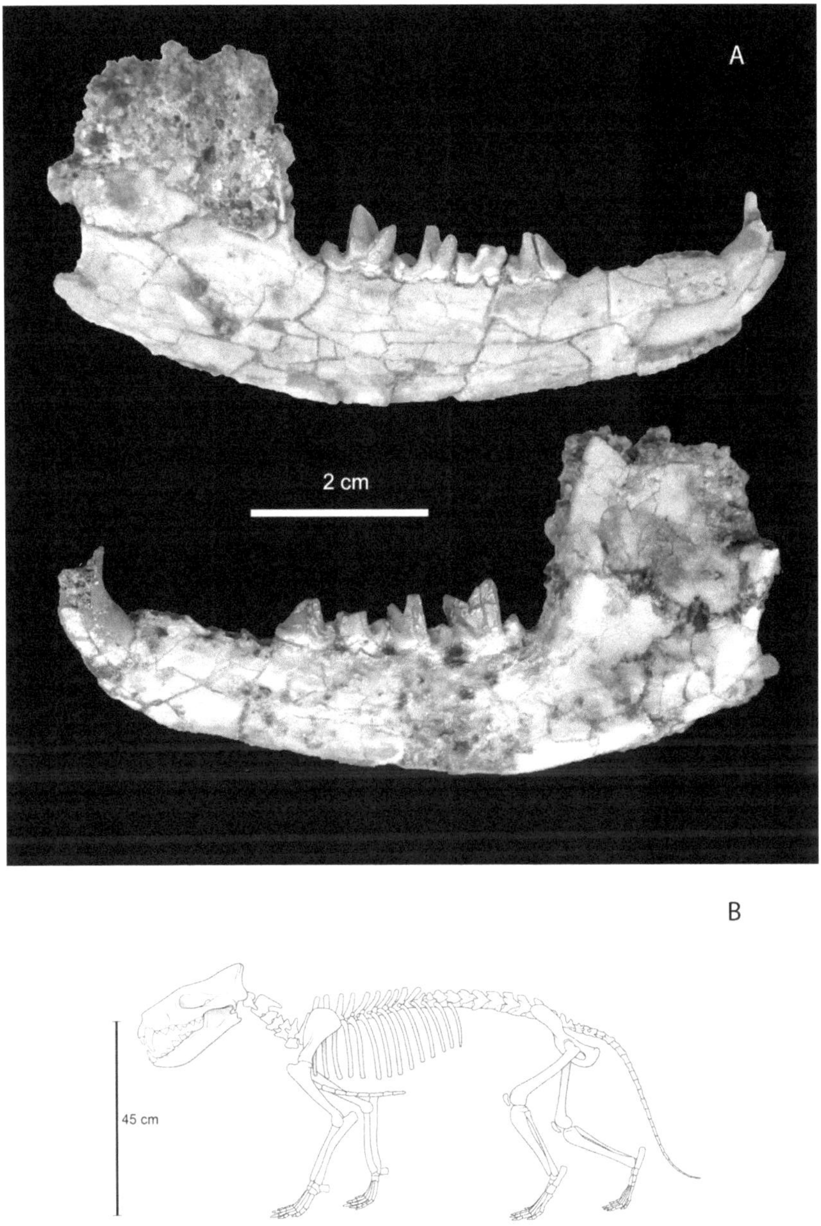

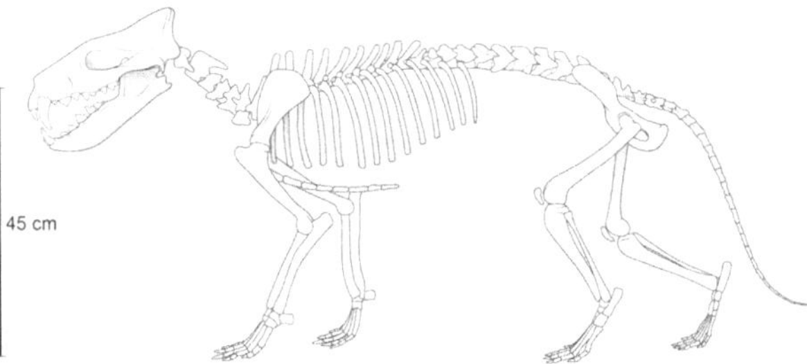

Figure 4.13. A. Side views of the lower jaw of the hyaenodont *Furodon crocheti* from the early–middle Eocene of Algeria, circa 48 Ma. *Furodon* was a stem representative and the earliest-known of the African hyainailourines. B. Reconstruction of the skeleton of a hyaenodont (*Hyaenodon*). From Solé et al. (2014); photos courtesy of F. Solé and R. Tabuce (**A**). Redrawn by A. Lethiers (**B**).

from Algeria (Gour Lazib, 48 Ma) is a small generalist terrestrial carnivorous mammal.

Apart from a possible Paleocene occurrence in Morocco, the earliest endemic African modern insectivores—that is, the earliest afrotherian tenrecoids—are found in the middle Eocene Namibian site of Black Crow, which is dated to about 43 Ma. They are represented by the genera *Nanogale, Diamantochloris*, and *Damarachloris*. *Damarachloris* and *Diamantochloris* are even related by Martin Pickford to the family Chrysochloridae that includes the living golden moles (Pickford, 2015c, 2018a, 2019). These mammals have the typical zalambdodont puncturing molars of the tenrecoids, although in a poorly advanced state with respect to the extant families, which might indicate they are stem tenrecoids. These Namibian tenrecoids are, however, paradoxically more evolved than the afrotherian insectivores described in the more recent Fayum sites of the late Eocene and early Oligocene.

The elephant shrews (macroscelideans), another group of afrotherian insectivores, are first found in the middle Eocene Tunisian site of Chambi with the species *Chambius kasserinensis* **(fig. 4.14)**, described in 1986 by Jean-Louis Hartenberger from the University of Montpellier. *Chambius kasserinensis* is related to the macroscelideans, but it is a very basal and distant species compared to the extant elephant shrews (sengis). For example, its teeth are brachyodont; that is, they have a low crown instead of the high crown of modern species. *Chambius kasserinensis* is probably a basal macroscelidean and the oldest-known member of the group. Intriguingly, it shows similarities in tooth and tarsal morphology to some European stem ungulates, such as the louisinid "condylarths," and to paenungulates, which led some paleontologists to consider that there could be a close relationship between elephant shrews and African ungulates. A surprising point to note is the much later appearance in Africa of these insectivorous afrotherians than the herbivorous afrotherians (the paenungulates). There is a difference of about 12 million years. This is another illustration of how heterogeneous and incomplete the African fossil record is.

The figurehead group in the rapid evolution of mammals in Africa during the middle Eocene is that of the dassies, the Hyracoidea. Hyracoids are

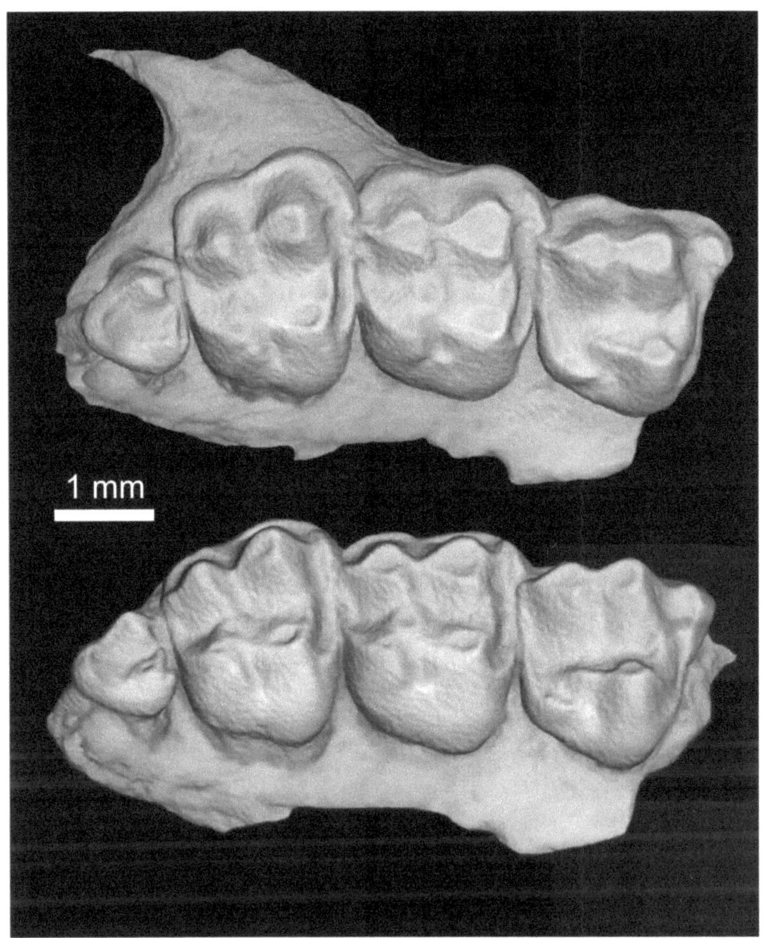

Figure 4.14. The earliest-known relative of the elephant shrews (macroscelideans): *Chambius kasserinensis*, from Chambi, Tunisia, early–middle Eocene, circa 48 Ma. The fossil seen here is the type specimen, a partial upper jaw with posterior teeth (top views; 3D digital images reconstructed from CT scans). From Tabuce (2018); image courtesy of R. Tabuce.

par excellence *the* endemic herbivorous mammals of Island Africa. They were the dominant herbivores in Island Africa in terms of abundance and adaptive and taxonomic diversity. They exploded in diversity for the first time in the Algerian sites of Gour Lazib (48 Ma), where they are represented by six genera and seven species from two families, the geniohyids and titanohyracids (Tabuce et al., 2011). They have varied morphologies and adaptations in locomotion and diet, and they have a wide range of body sizes, unknown today, with small groundhog-sized species of a few kilograms, such as *Microhyrax lavocati*, and species the dimension of a small rhino, such as "*Titanohyrax*" *mongereaui* (genus identification questionable) (body mass 700 kg). There are omnivorous, folivorous, arboreal, cursorial (runners), and terrestrial generalist species. These hyracoids illustrate most spectacularly the African mammal adaptive radiation. The Gour Lazib faunas (48 Ma) show that the basal radiation of hyracoids began at least as early as the middle Eocene (Adaci et al., 2007; Sudre, 1979; Tabuce et al., 2001). In these Algerian faunas, there are some hyperspecialized species that long remained enigmatic, such as *Helioseus insolitus* (Sudre, 1979; Tabuce, 2010), which has strange semicircular sawlike teeth; these very peculiar teeth, which are called "plagiaulacoid teeth" (large, bladelike, and serrated teeth), have evolved convergently in other extinct and unrelated groups of mammals, such as multituberculates, plesiadapiform primates, and even in marsupials. Paradoxically, the first radiation of the hyracoid group illustrated by fossils from the middle Eocene of Algeria indicates that the group goes back to a much earlier age and that it remains poorly known. The Gour Lazib hyracoids are currently under study by Rodolphe Tabuce.

In 2008, Martin Pickford reported the discovery of the skull of the embrithopod *Namatherium blackcrowense* from the Lutetian Black Crow site (ca. 43 Ma) in Namibia (Pickford et al., 2008) **(plate 12)**. At that time, the embrithopods were known in Africa only from the late Eocene and Oligocene genus *Arsinoitherium*. *Namatherium blackcrowense* is the only middle Eocene embrithopod known in Island Africa. This species is more or less contemporaneous with the embrithopods of the palaeoamasiid family (genus *Palaeoamasia*) that emigrated from Africa across the Tethys Sea into Anatolia and Romania. It is a good-sized mammal, with a body mass of about 350 kilograms, although smaller than *Arsinoitherium*. Among embrithopods it is most closely related to *Arsinoitherium* from the early Oligocene and belongs to the same family Arsinoitheriidae (Gheerbrant et al., 2025). *Namatherium blackcrowense* has a retracted nasal opening (nostrils pushed back), a large nasal cavity, and wide zygomatic arches for the attachment of powerful jaw muscles. *Namatherium* is, however, much less specialized than *Arsinoitherium*. For example, its teeth have a primitive low crown unlike *Arsinoitherium*. Upper molars are typically hyperdilambdodont with very wide W-like crests, as in all embrithopods, but in a more advanced stage than in *Stylolophus* and the palaeoamasiids: The W-like crests (ectoloph) are transversely elongated and begin to resemble the lophodont crests of *Arsinoitherium*, in which the tooth morphology is simplified and reduced to two long transverse crests. The palate is deep, as in *Arsinoitherium*, presumably for housing a large and prehensile tongue used for grasping leaves, fruit, and other plant foods while foraging. Although the broken roof of the skull in the only known specimen does not preserve nasal horns, some morphological details suggest that *Namatherium* had nasal horns like *Arsinoitherium* (Pickford et al., 2008).

Middle Eocene proboscideans are very poorly known. They are represented by a small, isolated molar discovered in 1953 in the Lutetian of Senegal (Gorodiski and Lavocat, 1953). This tooth was recently named *Saloumia gorodiskii*, in dedication to its discoverer, Alexandre Gorodiski (Tabuce et al., 2020). It is distinguished by a bunodont crushing morphology (i.e., teeth with low and swollen relief) and other traits that relate its proximity to *Moeritherium*. *Saloumia* would indeed be the earliest-known moerithere, although it cannot be ruled out that it is an older lateral lineage of proboscideans (Tabuce et al., 2020). The diversity and evolutionary history of proboscideans during the middle Eocene is obscured by a fossil gap in Africa (Tabuce et al., 2020) **(see fig. 2.6)**. In particular, the absence of proboscideans in the rich Gour Lazib sites (48 Ma) remains a mystery (could this be the effect of specialized ecology and environment?). Important future discoveries of proboscideans in the middle Eocene of Africa are to be expected. This is confirmed by the recent report of an elephantiform tooth from the Lutetian of Togo (Hautier et al., 2021) that is 10 million years older than the Fayum palaeomastodonts.

Sea cows (sirenians) are first known at the continental Chambi site on the basis of a fossil of an isolated petrosal skull bone (Benoit et al., 2013). Although fragmentary, this is an important discovery. It adds to the early diversity of sirenians in Africa since the middle Eocene, and it provides fossil evidence for the endemic African origin of the sea cow order.

Sirenians most likely first evolved in the African fresh waters and then colonized the Tethys Sea from the African shores with increasingly greater aquatic adaptations, such as the loss of hind limbs. It should be recalled that in the early and middle Eocene, Africa was occupied by huge inland lakes in the vast basins of the Sahara, Congo, and Somalia (Couvreur et al., 2021) **(see fig. 2.3).** These lakes were likely key ancient biotopes for the first aquatic specialization and early evolution of the sirenians. Apart from the continental Chambi site, early sea cows are known in Africa from Lutetian marine sites, the richest of which being those of Djebel Mokattam in the Fayum (Egypt). The African sirenians known from the Lutetian are remarkably diverse with five species, again testifying to their little-known ancient history. Among these species is an unidentified species of the early family Prorastomidae known from a few bones from the phosphates of Senegal (Hautier et al., 2012). All other species are more derived. They are represented by protosirenids (*Protosiren*) and "halitheriines," a paraphyletic group including *Eosiren, Eotherioides* that are successive stem groups of extant sea cows.

Rodents appeared in Africa during the Lutetian in North African sites (Algeria and Tunisia). They are represented by the endemic family "Zegdoumyidae" (*Glibia, Zegdoumys*), which is the stem group of the extant anomaluroids. Their first discovery was made in Algeria in 1994 by Monique Vianey-Liaud from the University of Montpellier (Vianey-Liaud et al., 1994). The "Zegdoumyidae" from the Algerian Glib Zegdou sites are among the most abundant and most diversified known mammals in the sites. About 100 dental remains are known (Marivaux et al., 2011). The family "Zegdoumyidae" is paraphyletic. It is a nonhomogeneous ancestral family of the anomaluroids, which represent one of the two major adaptive radiations of rodents in Island Africa, together with the hystricognath rodents. Zegdoumyids include the oldest African stem groups of the two extant endemic anomaluroid families: the flying scaly-tailed squirrels (Anomaluridae) and the flightless scaly-tailed squirrels (Zenkerellidae). They are the fossil evidence of the ancient African evolution of the anomaluroids (Marivaux et al., 2011; Vianey-Liaud et al., 1994), which is also indicated by molecular dating (e.g., Heritage et al., 2016). Zegdoumyids have a specialized and varied dental morphology. In particular, they document early stages of the presence of five crests on the teeth ("pentalophodonty") and of the specialized tooth enamel microstructure known in the anomalures. They underwent a limited early African radiation. It is represented by about 10 species and five genera from the Lutetian, such as *Glibia*, *Glibemys*, and *Zegdoumys*. This diversity, coupled with the early specializations of the group, suggests that the evolutionary history of the anomaluroid lineage dates back at least to the early Eocene, if not earlier (Heritage et al., 2016; Marivaux et al., "New and Primitive Species," 2014). The anomaluroids likely arrived in Island Africa following a dispersal from Eurasia, although their northern stem group is still not clearly identified.

Apart from the anomaluroids, the only other rodents known in Island Africa are the hystricognaths (also called hystricomorphs). Hystricognaths are represented in Africa by phiomorphs, such as the living cane rats, dassie-rats, and mole-rats. Hystricognaths make their first appearances in Africa during the late middle Eocene, in the Bartonian stage by 39.5 Ma, with the species *Protophiomys tunisiensis* from Djebel el Kébar (Marivaux et al., "New and Primitive Species," 2014), which is their basalmost stem group. No other African hystricognath are known in the middle Eocene. Early African hystricognaths, such as *Protophiomys tunisiensis*, probably arrived from Asia, following a dispersal across the Tethys Sea.

The first-known African strepsirrhine primates related to the living lemurs and lorises appeared in the middle Eocene at the Glib Zegdou and Chambi sites with two families, the azibiids and the djebelemurids. They are different from the strepsirrhines of the Eocene of Laurasia, which belong to a large, distinct radiation of stem groups represented by the extinct adapiforms. These early African strepsirrhines are represented by diminutive species, with body masses of around 100 grams, that belong to the genera *Algeripithecus* (Godinot and Mahboubi, 1994) and *Djebelemur* **(fig. 4.15)** (Hartenberger and Marandat, 1992). They are stem groups of modern strepsirrhines, such as lemurs and lorises, having evolved prior to the divergence of living lemurs (lemuriforms) and lorises (lorisiforms) (Marivaux et al., "Morphological Intermediate," 2014). In particular, they lack the specialized tooth comb that typifies modern lemurs and lorises.

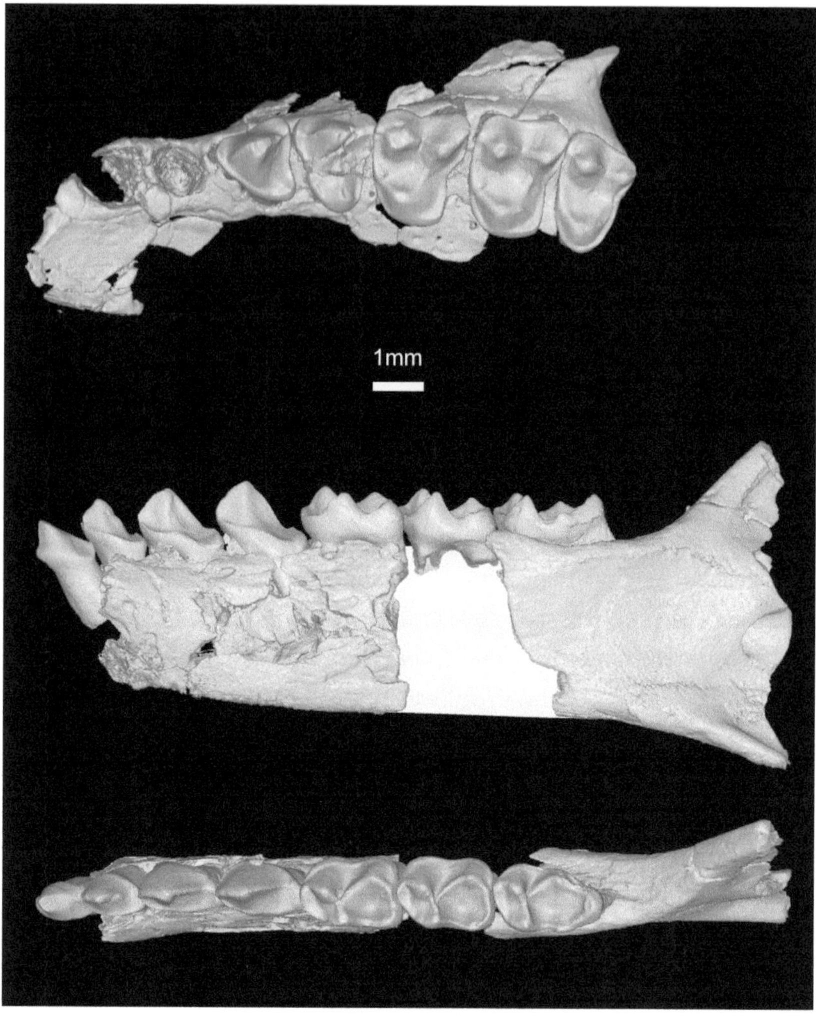

Figure 4.15. Upper and lower jaws with cheek teeth of the stem strepsirrhine primate *Djebelemur martinezi*, an early relative of both the lemurs and galagos (*top*, ventral view of maxilla; *middle and bottom*, internal and dorsal views of dentary) from Chambi, early–middle Eocene of Tunisia, circa 48 Ma; 3D digital image reconstructed from CT scans. Image courtesy of L. Marivaux.

The extinct adapiform strepsirrhines, which were successful in Laurasia during the Eocene, made occasional dispersals across the Tethys Sea, from Eurasia to Island Africa, without ever becoming permanently established in Africa. Their first record in Africa is reported from the

Lutetian of Namibia with the genus *Namadapis*, known from a lower jaw (Godinot et al., 2018). *Namadapis* is likely of European origin.

The anthropoid primates, which include monkeys and humans, appeared in Africa at the end of the middle Eocene with the ancestral species *Amamria tunisiensis* described by Marivaux in the Bartonian of the Tunisian site of Djebel el Kébar (41 Ma) (Marivaux et al., "Morphological Intermediate," 2014). *Amamria*, as important as it is, is known from only a single isolated molar **(plate 13 upper)**. It has a primitive morphology, intermediate between the Eocene Asian eosimiids, which are the first-known anthropoids, and the modern apelike African anthropoids (simiiforms). *Amamria* supports the view that the anthropoids colonized Africa from an ancient Asian cradle in the middle Eocene. Asia is now considered to be the center of origin of anthropoids. This is indicated by the eosimiids found in the early Eocene of China. The anthropoid primates arrived in Africa from Asia at the same time as the hystricognath rodents. These small mammals dispersed together across the Tethys Sea on the floating islands repeatedly generated between 41 and 40.5 Ma by the monsoons during the Middle Eocene Climatic Optimum (MECO) period (Beard et al., 2017; Scotese et al., 2021). The MECO floating islands are one of the ancient links in the origin of our hominid family.

Eocene African bats (chiropterans) were poorly known until recently. Our knowledge of them has greatly increased with new discoveries in the Lutetian sites of Glib Zegdou and Chambi, studied in the PhD thesis of Antony Ravel (Ravel et al., 2016). Ravel's thesis shows that bats suddenly appeared in Africa during the middle Eocene with a remarkable diversity, with 12 species in eight families of insectivorous microbats (microchiropterans). What is interesting is that these early African bats are mostly of modern groups, in contrast to the Laurasian bats of the same age, which are mostly stem groups (eochiropterans).

The early diversity of African bats results from the evolutionary radiation of three large modern extant groups, the Rhinolophoidea, Emballonuroidea, and Vespertilionoidea. Several species from Glib Zegdou and Chambi are thus the oldest known of the modern families of Hipposideridae (Old World leaf-nosed bats), Emballonuridae (sheath-tailed bats), and Vespertilionidae (common vesper, or simple nosed bats). The middle

Eocene bats of North Africa, in particular the families Hipposideridae (?*Palaeophyllophora*, *Hipposideros*), Emballonuridae (*Vespertiliavus*), and Necromantidae (*Necromantis*), show European relationships, especially with the bats from Quercy (France). They indicate middle Eocene dispersals of bats between Island Africa and Europe, and likely from North Africa to southern Europe. Several other microbat groups (Philisidae, and the Nycteridae *Khoufechia* and *Drakonycteris*) are probably of endemic African origin. A possible fruit bat (Megachiroptera) of the pteropodid family has been reported by Pickford (2018b) in the Lutetian of the Black Crow site in Namibia. It remains poorly known (a humerus and a possible tooth). This is a medium to large chiropteran in general, and a small species compared with fruit bats. However, it is important as it might be the earliest-known megachiropteran. The species *Tanzanycteris mannardi* from the Lutetian of Tanzania (Gunnell et al., 2003) is one of the few stem bats (eochiropterans) known in Africa. It is a witness to the cosmopolitan distribution of eochiropterans.

The first representatives in Africa of the whale group, the Cetacea, are archaeocetes (i.e., archaic toothed whales). They are large mammals (>100 kg) with elongated skulls that specialized very early. They displayed increasing aquatic adaptation. Archaeocetes originated on the Indian ocean shores of the Tethys in the early Eocene from an artiodactyl ancestor shared with hippopotamuses, and they subsequently rapidly colonized the entire Tethys Sea, from east to west. They appear in Africa during the middle Eocene in several sites found on the Tethyan and Atlantic shores of the continent. Some of these sites are located far inland, which reflects the extension of the epicontinental seas into the African continent. The first African archaeocetes are represented by semiaquatic protocetids, such as *Protocetus* and *Pappocetus* (Lutetian–Bartonian, 45–38 Ma). They are followed by the more specialized and fully aquatic basilosaurs *Dorudon* and *Eocetus* (Bartonian). About 10 species of early African archaeocetes have been identified. The richest sites are those of Egypt. Latest discoveries of archaeocetes in Africa were reported by Zouhri and Gingerich in the Bartonian (41–38 Ma) of the Moroccan Sahara at the rich Gueran sites (Zouhri et al., 2021).

Late Eocene, 38–34 Ma

The late Eocene (Priabonian, 38–34 Ma) is a period that is fairly well represented by fossil mammals in Africa. There are about 50 known late Eocene mammal species, as many as in the middle Eocene, which is a longer period (10 million years). The most important fossil sites are those of Fayum-BQ2 (Birket Qarun, Egypt), Dur at Talah (Libya), and Bir El Ater (Algeria). Several important new appearances are recorded at this time in Island Africa, including artiodactyls (anthracotheres), lorisiform primates (galagos) and possibly tarsiiform primates (tarsiers), new families of anomaluroids and phiomorphs rodents and bats, and important new genera such as the embrithopod *Arsinoitherium* and the proboscidean *Moeritherium*. All evolved in situ as native African lineages except for the anthracotheres, which were immigrants from Eurasia. The late Eocene is part of a period of a continuous evolutionary radiation in Africa. The only notable extinctions are those of the stem bats and the toothed whales of the protocetid family.

The ongoing African evolution during the Eocene, without marked faunal breaks, is well illustrated by the hyaenodont carnivorous mammals. The two hyainailourid African lineages known earlier in the middle Eocene (i.e., the Hyainailourinae and Teratodontinae) are present in the late Eocene. An additional lineage, the apterodontines (*Apterodon*), appears and is present in abundance in most late Eocene African localities. Apterodontines are specialized semiaquatic hyainailouroids. According to the robust morphology of their teeth, they fed on prey with hard shells, such as crustaceans. The teratodontine *Masrasector* made its first appearance at Bir El Ater (Algeria). The Bir El Ater species is close to *Masrasector ligabuei*, known from the more recent Omani site of Taqah (34 Ma).

Among African ungulates, hyracoids are poorly known in the late Eocene in contrast with older faunas from the middle Eocene. They are only known by two small species, the geniohyid *Bunohyrax matsumotoi* from Bir El Ater (Algeria) and *Dimaitherium patnaiki* from Fayum-BQ2 (Egypt). *Dimaitherium patnaiki* is the oldest hyracoid found in the Fayum sites. *Dimaitherium* is known by a good collection of fossils, including skulls and limb bones (Barrow et al., 2010). These fossils show that *Dimaitherium* was a small, agile climber and able to move fast, without being specialized in running. *Dimaitherium* has a primitive hyrax morphology—for example, in its simple premolars. It belongs to a basal hyracoid branch that succeeds the older genera *Seggeurius* and *Microhyrax*. This early basal lineage probably also includes *Namahyrax* from the middle Eocene site of Black Crow (Pickford et al., 2008), which shares close affinity with *Dimaitherium*.

Late Eocene proboscideans are well represented at the sites of Dur at Talah (Libya) and Fayum–Qasr El Sagha (Egypt). They include the stem barytheriid family characterized by lophodont cutting teeth. The huge species *Barytherium grave*, of elephantine size (body mass two metric tons; shoulder height over 180–200 cm), is the most famous barytheriid. *Barytherium* is known from few fossils found at the beginning of the 20th century in the Fayum, and from much more abundant fossils found in the Dur at Talah sites. Its skeleton was reconstructed in 2005 in Cyrille Delmer's PhD thesis (**fig. 4.16**). *Barytherium grave* was a massive animal with a huge skull. Its skeleton resembles earlier stem proboscideans, such as *Numidotherium* (for instance, in the limbs that are flexed). It has plantigrade hand and feet, and the hand is noticeably flexible. The plantigrade and flexible manus, together with a peculiar morphology of the humerus, indicates a capacity of its anterior limbs for foraging for food. In addition, the manus of *Barytherium grave* is broad, and the limbs are very inclined (as in rhinos for instance) resulting in a low center of gravity of the body. This indicates that *Barytherium grave* lived in a closed and humid environment with soft ground, such as marshes, mangroves, riverbanks, and seashores (Delmer, 2005). Geochemistry of the teeth of *Barytherium* has supported the hypothesis that *Barytherium* grazed on freshwater aquatic plants in swamps and rivers (Liu et al., 2008). *Barytherium grave* is associated in the Dur at Talah sites with another, smaller proboscidean, *Arcanotherium savagei* (Delmer, 2009), which remains poorly known (mostly teeth and jaws).

The other group of proboscideans known from the late Eocene of Africa is the family Moeritheriidae. Moeritheriids are known from the Eocene to the Oligocene, but the richest sites are those of the late Eocene marine Qasr

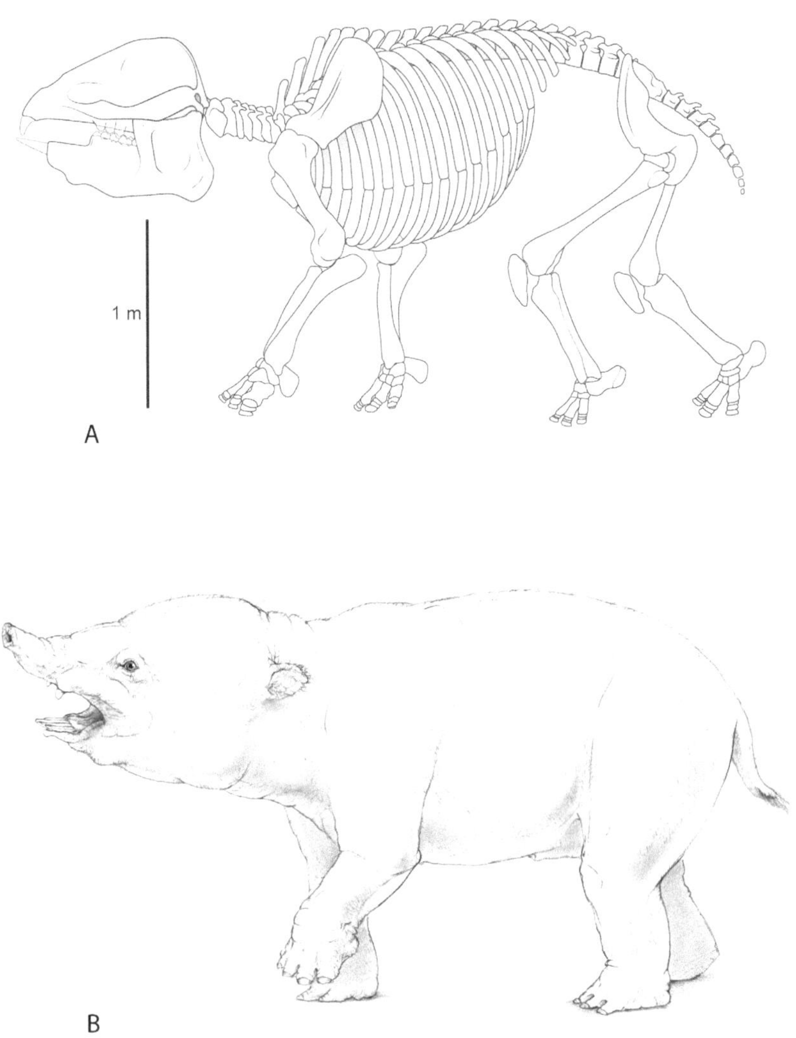

Figure 4.16. The large stem proboscidean *Barytherium grave*, discovered in the late Eocene fossil sites (37–34 Ma) of the Fayum (Egypt) and Dur at Talah (Libya). **A.** Skeletal reconstruction (length about 3.5 m). **B.** Life restoration. Redrawn by A. Lethiers from Delmer (2005) (**A**). Drawing by C. Letenneur (**B**).

El Sagha Formation (35 Ma) of the Fayum. Elwyn L. Simons found in these levels an almost complete skeleton of *Moeritherium lyonsi*, exhibited in the Cairo Geological Museum (Simons, 1964) **(fig. 4.17)**. It was an animal with the dimensions of a hog or a tapir (shoulder height of 70 cm; body mass of

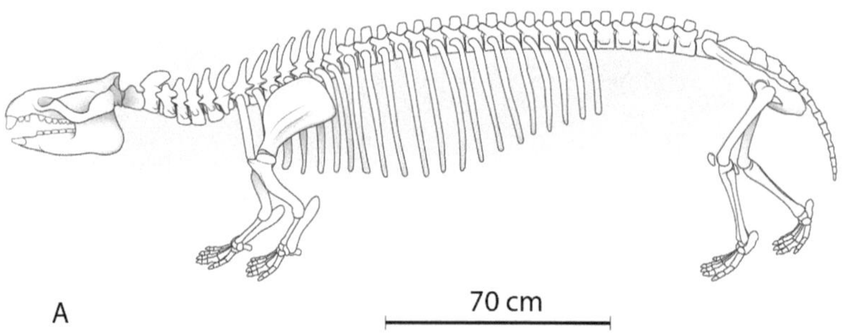

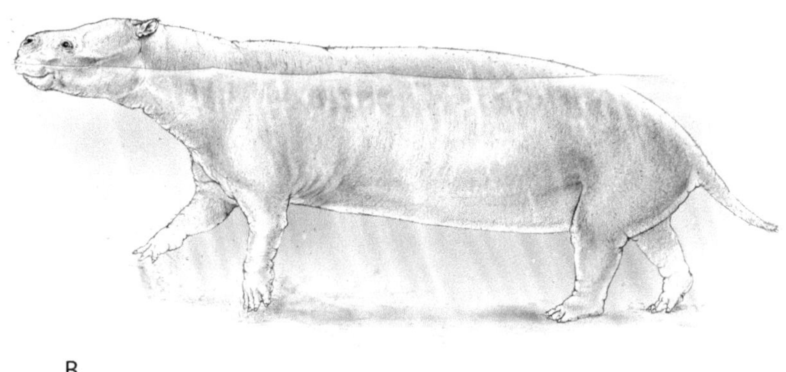

Figure 4.17. The stem proboscidean *Moeritherium lyonsi* from the late Eocene of the Fayum (Qasr El Sagha Formation, 35 Ma). *Moeritherium* was a peculiar specialized early relative of the elephants. It was adapted to a semiaquatic life with a distinctive elongated body shape and short legs. Its body was 230 centimeters in length and only 70 centimeters at the shoulders (body mass was 235 kg). **A.** Drawing of the mounted skeleton of *Moeritherium lyonsi* from the late Eocene of the Fayum (Qasr El Sagha Formation, 35 Ma), exhibited in the Cairo Geological Museum (reconstruction made by Grant E. Meyer and Arnold Lewis). **B.** Life restoration of *Moeritherium*. Drawing by A. Lethiers (**A**). Drawing by C. Letenneur (**B**).

230 kg). The skeleton shows the animal had a strange body for a proboscidean, in the shape of an elongated barrel on short legs. It remarkably resembles stem sea cows, such as *Pezosiren*, from the Eocene of Jamaica. In addition, *Moeritherium lyonsi* has a peculiar elongated and tubular skull, with the orbit positioned very high and anteriorly. This peculiar morphol-

ogy is linked to a specialized semiaquatic lifestyle. The detailed study of a *Moeritherium lyonsi* skull from Dur at Talah (Tassy, 1981) has shown that moeritheres are closely related to modern proboscideans, the elephantiforms. For example, *Moeritherium* shares with elephantiforms the beginning of a lighter skull (i.e., with an early pneumatization of the bones), an enlarged second upper incisor, and bunolophodont molars with an incipient third loph that forms an additional cutting crest. Moeritheres were less specialized than elephantiforms. For example, they retained canines, and they did not have tusks and a trunk, as indicated by the nostrils located anteriorly on the skull. The first Eocene moeritheres were plesiomorphic, but they quickly specialized and became amphibious. *Moeritherium lyonsi* had a lifestyle and locomotion close to that of the hippopotamus. It ate aquatic plants and, particularly, mangrove plants, as indicated by studies of the isotopic composition of its teeth (Clementz et al., 2008). The Eocene moeritheres have an intermediate morphology between barytheriids and elephantiforms but are more derived than *Arcanotherium*. This is shown by the small Bir El Ater species *Moeritherium chehbeurameuri*, which has lophodont cutting teeth in the process of becoming bunolophodont: The tubercles of the molar teeth are swollen and multiplied, and the cutting crests are weakened, trending toward a more crushing function. This species can be qualified as a "missing link" between the basal folivorous proboscideans of the Eocene and the modern crushing proboscideans of more recent age. It has also been reported from Dur at Talah, in Libya (Jaeger et al., 2012). Another small plesiomorphic, *Moeritherium*, is reported from the late Eocene of Mali. Other species of *Moeritherium*, such as *M. gracile*, have been mentioned in the Fayum–Qasr El Sagha sites, but they remain poorly discriminated from *Moeritherium lyonsi*.

The embrithopods are poorly known in the late Eocene. Their fossils, found at Dur at Talah in Libya and at the Fayum (Qasr El Sagha) in Egypt, are few and fragmentary. They are nevertheless important because they testify to the first appearance of the classic and most derived embrithopod genus, *Arsinoitherium*. The late Eocene species of *Arsinoitherium* found in Egypt and Libya remains unidentified.

The sea cows (sirenians) known in the late Eocene of Africa mostly belong to the "halitheriines," a paraphyletic assemblage of the closest stem groups to extant sea cows, which comprise manatees and dugongs. They are not very diversified and are represented by two species—which are possibly synonymous—of the genus *Eosiren* already known in the middle Eocene. Abundant fossils of a late Eocene sirenian comprising skulls and postcranial bones have been found in the Fayum (Qasr El Sagha) and in Libya (Dur at Talah) and Morocco (Dakhla). The species *Protosiren smithae* from Wadi Hitan (Egypt) is the last-known protosirenid in Africa.

The elephant shrews or sengis (macroscelideans, insectivorous afrotherians) known from the late Eocene are slightly more diverse than in the middle Eocene. They are known from the Bir El Ater (Algeria) and Dur at Talah (Libya) sites by isolated teeth of three species of two genera, *Nementchatherium* and *Eotmantsoius*, all belonging to the stem group Herodotiinae. The species *Nementchatherium rathbuni* from Dur at Talah is clearly more derived than *Nementchatherium senarhense* from Bir El Ater, supporting the more recent age of the former locality.

In the late Eocene sites of Africa, we find anomaluroid rodents that are descendants of the Lutetian zegdoumyids. They are represented by the paraphyletic (unnatural) stem family "Nementchamyidae," which includes the genera *Nementchamys* (Bir El Ater) and *Kabirmys* (Dur at Talah, Fayum-BQ2). They are more evolved than the zegdoumyids and closer to the extant anomalurid family of the African flying squirrels. *Nementchamys* was the first anomaluroid discovered in the Paleogene of Africa (Jaeger et al., 1985). The extant anomalurid family of the flying squirrels appears at this late Eocene time with the genus *Shazurus*, discovered in the Fayum-BQ2 locality.

The hystricognath rodents that appeared in Africa at the end of the middle Eocene (40–39 Ma) became well diversified in the late Eocene, with the presence of four genera (*Protophiomys, Talahphiomys, Waslamys,* and *Phiomys*). They are successive stem groups of the extant African phiomorphs, which include the cane rats, mole-rats, and dassie-rats. The basal-most African hystricognaths from the late Eocene are *Talahphiomys* from Dur at Talah and *Waslamys* from the Fayum-BQ2 site. More specialized late Eocene stem phiomorphs also occur, such as the phiomyids *"Phiomys" hammudai* (genus identification questionable) from Libya (Dur at Talah) and *Protophiomys* from Algeria (Bir El Ater). The diversity of these stem hystricognaths testifies to an ancient history in Island Africa that goes back well before the late Eocene.

Strepsirrhine primates of the extinct Laurasian adapiform group were present in Africa in the late Eocene. They colonized Africa from Eurasia across the Tethys during the middle Eocene (Seiffert et al., 2018). Two species of caenopithecine adapids, *Afradapis longicristatus* and *Masradapis tahai*, were discovered in the Fayum-BQ2 site by Erik Seiffert in 2009 and 2018. They illustrate an unsuspected diversity of the African adapids. But, most remarkably, these African adapids show convergent specializations with the anthropoids. In *Afradapis*, for example, we observe the presence of a fused mandibular symphysis, a very deep mandible, a canine with a sharp wear facet, a reduced dental formula with only two premolars, oblique premolars, and a large molar hypocone (the fourth tubercle of upper molar). These morphological convergences indicate that there was competition between adapids and anthropoids in the same ecological niches of Island Africa for a short time. However, the adapids remained short-lived in Africa, with no descendants; they had no chance against the competition and evolutionary success of the anthropoids in Island Africa. *Afradapis*, with a body mass of about three kilograms, was the largest nonanthropoid primate found in Island Africa. It was a highly specialized primate with lophodont teeth adapted to a folivorous (leaf-eating) diet.

It was in the late Eocene that the first species of the extant group of lorises and galagos (lorisiforms) appeared in Africa **(fig. 4.18)**. They are represented by *Karanisia clarki* and *Saharagalago misrensis* from the Fayum-BQ2 site (37 Ma). *Saharagalago* is seen either as the earliest galago or as a stem lorisiform. *Karanisia* and *Saharagalago* are known from fossil jaws with a tooth comb, unlike the older, middle Eocene, strepsirrhines. They are the oldest-known strepsirrhines with such teeth. The tooth comb is a specialized tool used for feeding on tree gum (by scraping it) and for grooming the coat, in the absence of claws (instead they have fingernails). A recent study of the *Karanisia* fossils has confirmed that it used its anterior teeth to obtain and ingest gums and saps (López-Torres et al., 2020).

The presence of afrotarsiids in the late Eocene of Africa was reported with the discovery of isolated teeth of *Afrotarsius libycus* from of Dur at

Figure 4.18. An extant African strepsirrhine primate, the diminutive bush baby *Galago* (body mass 200–500 g). The bush babies, like other lorisiforms, are nocturnal primates. Drawing by C. Letenneur.

Talah, in Libya (Jaeger et al., "Late Middle Eocene," 2010). *Afrotarsius* was first discovered from the Oligocene of the Fayum. The relationships of the afrotarsiids are disputed (see chapter 5).

After their very first appearance in the Bartonian of Djebel el Kébar (41 Ma), the earliest African anthropoids are found in the late Eocene (37 Ma) of Algeria, Libya, and Egypt. They were first identified in 1988 by Louis de Bonis (University of Poitiers) at the Bir El Ater site, with a single tooth of *Biretia piveteaui* (Bonis et al., 1988). Since then, more complete fossils of *Biretia* have been found in the Fayum (Seiffert et al., 2005) **(fig. 4.19)**, and we currently know three species of this genus. *Biretia* is a small anthropoid (body mass of about 300 g) with bunodont teeth and a strong hypocone, which is the fourth upper molar tubercle specialized for crushing the fruits on which these primates fed. The morphology of the

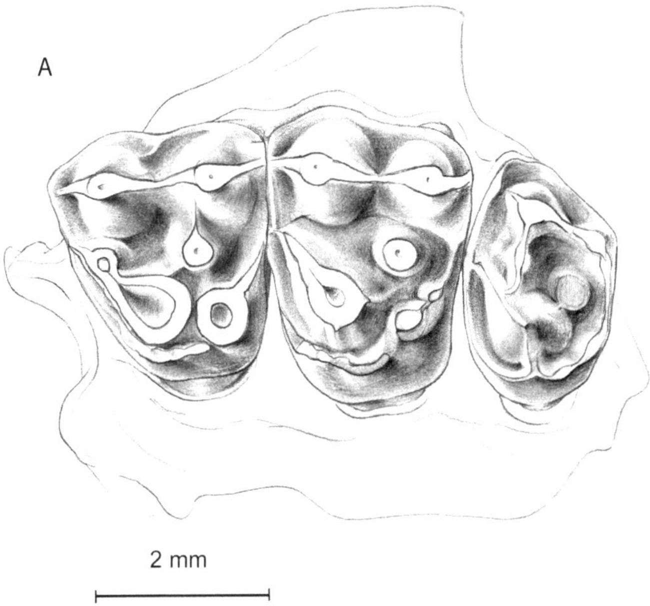

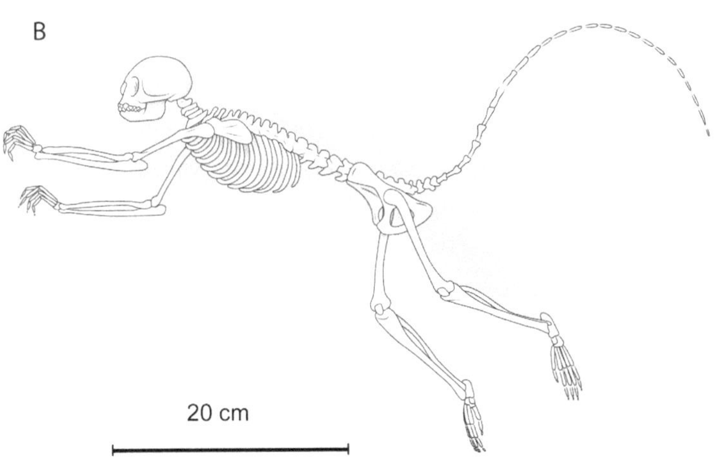

Figure 4.19. Following the middle Eocene genus *Amamria* (**see also plate 13 upper**), the late Eocene genus *Biretia* is the earliest-known anthropoid in Africa. This early parapithecid has been found in 37-million-year-old levels in Algeria (Bir El Ater site) and in the Fayum site of BQ2 (Egypt). **A.** Fragment of upper jaw with molars found in the Fayum-BQ2 locality. **B.** Reconstruction of the skeleton of a parapithecid (*Apidium*; body mass 1.6 kg). Drawing by C. Letenneur from Seiffert (2005, fig. 2) (**A**). Redrawn by A. Lethiers from Fleagle (2013) (**B**).

teeth and jaws of *Biretia* is similar to the classical and more recent Fayum parapithecids, but it is more primitive in the unfused mandibular symphysis, for example. Fused mandibular symphysis evolved in the parapithecids by convergence with the catarrhines. *Biretia* is more evolved than *Amamria* from the Bartonian of Tunisia, and it is the oldest parapithecid related to the root of the group that includes all extant platyrrhines (New World monkeys) and catarrhines (Old World monkeys and apes). Another stem anthropoid of obscure relationship was reported by Seiffert from the Fayum-BQ2 site (Seiffert et al., 2010). It is called *Nosmips*, an anagram of the name of the famous American paleontologist G. G. Simpson. It is a much more frequent primate than *Biretia* in the Fayum-BQ2 site, being represented by numerous fossil jaws and bones. It is a fairly large species with strangely specialized morphology (especially for lower premolars and molars) that belongs to an old, enigmatic African lineage. Finally, *Talahpithecus*, although poorly known by a few isolated teeth from Libya (Jaeger et al., "Late Middle Eocene," 2010), represents an important, more derived anthropoid. It has been identified as the earliest oligopithecid and catarrhine, but this is debated.

All known African bats (microchiropterans) from the late Eocene come from the Fayum-BQ2 locality (37 Ma). Five species belonging to four modern families have been reported. With a body mass of about 80 grams, *Witwatia* was a large microbat. It belongs to the extinct family Philisidae (in the extant vespertilionoid group), already known from the Lutetian of Tunisia. The philisids are the most frequent bats found in Island Africa during the Paleogene; they have been found in Tunisia, Egypt, Libya, and Oman. The other three genera and their families—Myzopodidae, Aegyptonycteridae, and Rhinopomatidae—first appear in the late Eocene locality of the Fayum-BQ2. *Phasmatonycteris butleri* is the oldest representative of the family of sucker-footed bats from Madagascar (myzopodids). They are insectivorous bats that have specialized adhesive discs (like suction cups) on their hands and feet to move on the smooth surfaces of large leaves, like those of the Malagasy traveler tree. *Phasmatonycteris* supports the African origin of this family, which is today endemic to Madagascar (Gunnell et al., 2014). The myzopodid family itself belongs to a southern group of bats, the Noctilionoidea. *Aegyptonycteris* (family

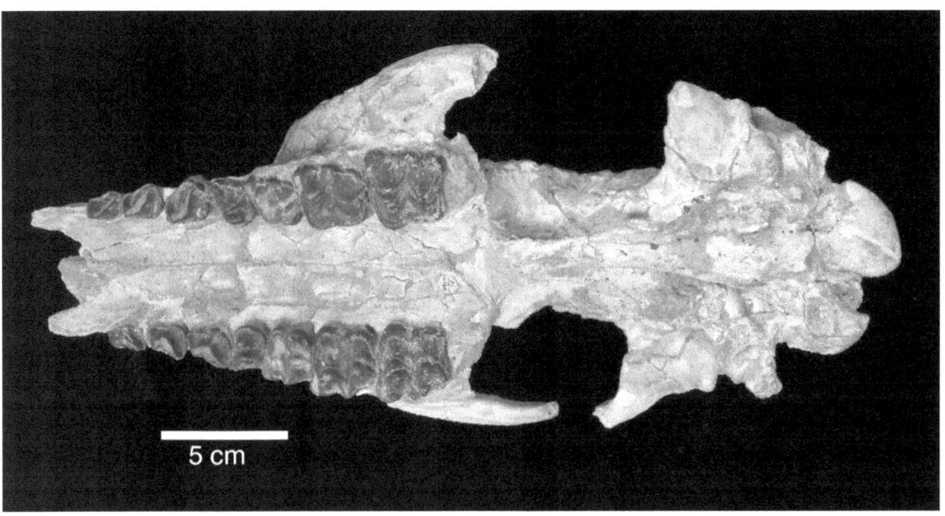

Figure 4.20. Skull of an African Paleogene anthracothere (Artiodactyla), *Bothriogenys fraasi*, from the early Oligocene (Jebel Qatrani Formation, upper sequence) of the Fayum (holotype NHMUK PV M 10186 from the Natural History Museum collection). Anthracotheres are among the most abundant mammals found in the Fayum. They were the first terrestrial artiodactyls to arrive in Africa. Image copyright The Trustees of the Natural History Museum, London, https://data.nhm.ac.uk/media/0418eebe-5ceb-46ec-b0ee-62a8adcb8432. Licensed under CC BY 4.0.

Aegyptonycteridae), known by a maxilla, was a large bat, close in size to the living fruit bats. It was the first African omnivorous bat, besides a poorly known possible fruit bat from the middle Eocene of Black Crow. It fed on insects, small vertebrates, and plants. Stem bats, and in particular the middle Eocene eochiropterans, are in fact more strictly "animalivorous" or insectivorous. They seem to have disappeared in the late Eocene in Africa.

Nonmarine artiodactyls (i.e., artiodactyls excluding cetaceans), which include pigs, hippopotamuses, and ruminants, first occurred in Africa in the late Eocene (Priabonian), around 37 Ma, with the anthracotheres (family Anthracotheriidae) **(fig. 4.20)**. This is the first euungulate group to colonize Africa from Laurasia, and it was the only one present in Island Africa. The late Eocene African anthracotheres are represented by an unidentified species of the basal genus *Bothriogenys* from the sites of Bir El Ater and Fayum–Qasr El Sagha. However, the

Bir El Ater fossils remain undescribed, and their identification needs to be confirmed. Anthracotheres originated in Asia, where they are known since the middle Eocene. They made an important evolutionary radiation from the Eocene to the Miocene throughout Laurasia (Eurasia and North America). The presence of the genus *Bothriogenys* in the late Eocene of Africa results from a dispersal from Asia at the end of the middle Eocene. Anthracotheres were large herbivores with robust skeletons. Their skeletal morphology resembled hippos, and the sedimentary deposits in which they have been found support the conclusion that they were semiaquatic. Their semiaquatic way of life probably explains their early arrival in Africa before all other Eurasian euungulates (see, for example, Lihoreau et al., 2015). The geochemistry of the teeth of *Bothriogenys* (stable isotope composition) confirms that it was a semiaquatic animal that ate aquatic plants (Clementz et al. 2008).

In the late Eocene, the large archaeocetes of the basilosaurid family flourished at the expense of the protocetids, which disappeared. Six genera of basilosaurs have been recorded in three late Eocene African localities: Dakhla (Morocco), Fayum–Qasr El Sagha, and especially the famous "Valley of the Whales" of Wadi Hitan in Egypt (Birket Qarun Formation) **(plate 13 lower)**. The two most famous species are *Basilosaurus isis* **(fig. 4.21)** and *Dorudon atrox*. The *Dorudon* lineage was already present in the middle Eocene. Basilosaurs were strictly marine predators; they were large, reaching 15 meters in length. They had an elongated, serpentine body shape, without hind limbs, and are characterized by the presence of many vertebrae. They had fearsome jaws armed with large teeth specialized as sharp blades resembling those of sharks. They rapidly colonized most of the Eocene seas.

Oligocene—climax of African endemism

The Eocene–Oligocene transition, around 34 Ma, is characterized in Laurasian continents by a faunal revolution called the "Grande Coupure" (the "great cut"), which was related to a major climatic change. A climatic cooling led to the development of glaciations for the first time in the

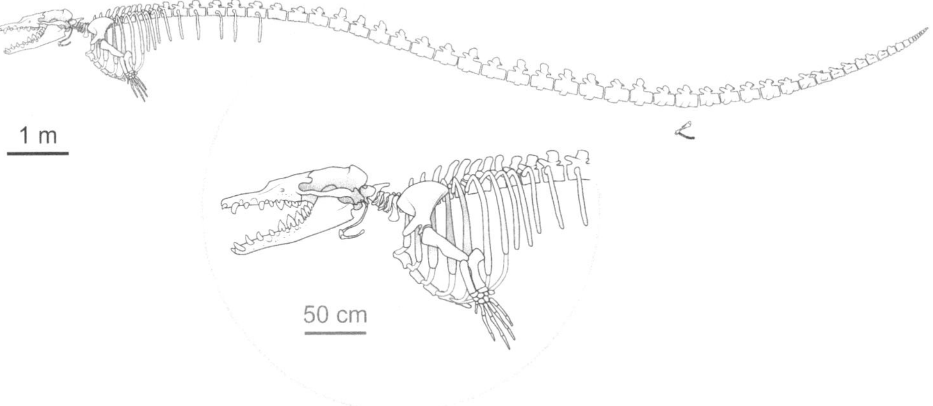

Figure 4.21. Reconstruction of the skeleton of *Basilosaurus isis* (whole skeleton, *above*; focus on the skull and forequarters, *below*) from Wadi Al-Hitan. *Basilosaurus* was a gigantic and fearsome marine top predator that was widespread in the Tethys Sea. Drawing by A. Lethiers after Voss et al. (2019).

Cenozoic, and to a drop in sea level. This resulted in a major extinction crisis in Laurasia and the evolutionary radiation of new taxa of Asia origin. In Africa, in contrast, it was a period of evolutionary continuity for most mammals. There were some extinctions, but they were on a much smaller scale. Interestingly, these extinctions especially concerned early marine mammal groups, such as the toothed whales (archaeocetes) and the pro-rastomid sea cows (sirenians). The period was nevertheless marked in Africa by the rapid and significant evolution of several endemic groups, leading to the emergence of new modern lineages, such as the elephantine-like proboscideans (elephantiforms) and the catarrhine primates.

The Oligocene was the climax of African endemic evolution. The African groups were flourishing. They reached their maximum diversity and abundance in all sites, especially in the Rupelian. Some endemic species were spectacular, especially among the paenungulates (i.e., the proboscideans, embrithopods, and hyracoids). African primates and rodents abounded. The hyaenodont diversity was maximal. There were specialized groups unknown elsewhere, such as the enigmatic ptolemaiids.

Some new appearances are recorded during the Oligocene, such as pangolins (Pholidota), hamster rodents (cricetids), rare true carnivores, and catarrhine primates. The pangolins, carnivores, and cricetids were immigrants, whereas other mammals evolved locally in Island Africa.

Early Oligocene, Rupelian, 34–28 Ma

The greatest number of mammals known in Island Africa is found in the Rupelian, with about 135 identified species; this is nearly half (44%) of all known Paleogene species in Africa. The Rupelian mammals have been found in 10 major African localities, but most of them come from the Fayum (Egypt) in the Jebel Qatrani Formation and, to a lesser extent, from Taqah (Oman). Only three Oligocene sites are in known sub-Saharan regions of Africa: Malembe (Angola), Rukwa (Tanzania), and Sperrgebiet (Namibia). However, their age is not well established. This is particularly true for the sites reported by Pickford's team in Namibia, such as Eocliff, Eoridge, Silica North, and Silica South. These Namibian sites were initially dated from the Eocene (see Pickford, 2015b, and references therein), but this is controversial. Several researchers consider these sites to be from later than the Eocene, or even later than the Paleogene (Asher, 2019; Marivaux et al., "New and Primitive Species," 2014; Sallam and Seiffert, 2016). The difficulty in the dating of these sites is that they are strictly continental and there is no constraining biostratigraphic or geochronological data; their dating is primarily based on the content and evolutionary stage of their mammal assemblage. Nevertheless, the Eoridge and Eocliff sites have provided some faunal data suggesting an Oligocene age. Eoridge, in particular, has yielded the anthracothere *Bothriogenys gorringei* (Pickford, 2015a), known in the early Oligocene Jebel Qatrani Formation of the Fayum, and the hyracoid "*Rupestrohyrax lacustris*" (Pickford, 2015e), which is a possible synonym of *Titanohyrax angustidens* (Asher, 2019), also known from the Jebel Qatrani Formation (upper levels).

In the Fayum, there are nearly 90 local sites dated from 34 to 29 Ma in the Jebel Qatrani Formation. However, they are not equally fossiliferous: The most abundant fossils by far come from the sites of L41 (ca. 34 Ma) and Quarries A–B (33 Ma), E, G–V (31 Ma), and I–M (30–29 Ma).

Apart from the early Eocene genus *Kasserinotherium*, which is of dubious relationships, marsupials first appeared in Island Africa in the early Oligocene. A single species, *Peratherium africanum*, was discovered in the Fayum (Quarry M, 30 Ma, Jebel Qatrani Formation) and Taqah (Oman). It belongs to the European family Herpetotheriidae. This family was initially considered to be related to the living opossum-like marsupials, but the latest studies indicate it is a stem group of all extant marsupials. For Hooker and colleagues (2008), the close relationship of *Peratherium africanum* to *Peratherium lavergnense* from the late Eocene (Priabonian) of Europe indicates that the herpetotheriids arrived in Island Africa at the Eocene–Oligocene transition, by 34 Ma. Marsupials are among the mammals with the shortest history in the continent. If we disregard some doubtful Eocene identifications such as the upper molar of *Kasserinotherium*, African marsupials are indeed known only in the early Oligocene, with the species of Fayum and Taqah, and in the middle Miocene of Uganda with the genus *Morotodon* (Crespo et al., 2022).

Another discrete and ephemeral group of mammals in Africa, the ptolemaiidans (order Ptolemaiida), appears in the Oligocene in Africa. It is only known in Africa, and it disappeared soon afterward during the Miocene, around 12.5 Ma. Ptolemaiidans are small and medium-sized insectivorous mammals, the largest reaching the size of a fox. They have only been found in the Oligocene of the Fayum with the Ptolemaiidae family and in the Miocene of Kenya with the Kelbidae family, a derived family of large ptolemaiidans. Their fossils, known since the beginning of the 20th century, are rare (about 50 specimens from the Fayum), and they are mostly jaw fragments. As a result, we don't know what they looked like.

The family Ptolemaiidae is restricted to the Oligocene. Five species in four genera have been described from three levels (Quarries A, V, I) of the Fayum dated from 34 to 29 Ma. The richest site is the Quarry V (31 Ma), where an exceptional skull of *Ptolemaia* was found. This site represents a peculiar depositional environment and habitat compared to all others in the Fayum, which would explain the abundance of ptolemaiids, which are otherwise rare. *Ptolemaia* had peculiarly specialized teeth, with a high crown (beginning of hypsodonty) and molars that became smaller toward the back of the jaw. This indicates an insectivorous mammal occupying a particular ecological niche: Its diet included abrasive foods, and it may have been a colonial insect feeder. *Qarunavus* is the most basal ptolemaiid. It represents the stem ptolemaiids from which two major lineages of ptolemaiids evolved (Bown and Simons, 1987). One these two lineages includes insectivorous ptolemaiids, such as *Ptolemaia*, and the other includes carnivorous ptolemaiids, such as *Cleopatrodon*. The relationship of the ptolemaiids to other mammals remains enigmatic. They have been related to archaic insectivore groups, such as the pantolestids, and to the enigmatic Eocene European paroxyclaenid family. More recently, a relationship with aardvarks and afrotherians has been considered (Kampouridis et al., 2023; Seiffert, 2007), because they share similar peculiar high and tubular teeth. However, an evolutionary convergence between these groups cannot be ruled out, and we must wait for new fossil discoveries to clarify the relationship of the ptolemaiidans within placentals. In any case, the ptolemaiids represent a small radiation of insectivorous mammals, unique to Africa, that went extinct after the middle Miocene. Their origin is probably ancient in Africa, but it is unknown because of the lack of fossils.

The African carnivorous mammals of the hyaenodont group were at their peak in the Oligocene, following an earlier long evolutionary history in Island Africa. The Eocene diversification of the hyainailouroids increased again during the Oligocene. There are about 15 African species (eight genera) belonging to three hyainailouroid subfamilies (Hyainailourinae, Teratodontinae, and Apterodontinae) in the Rupelian. These Oligocene African hyainailouroids varied in size and adaptations. The most abundant were *Apterodon* (Apterodontinae) and *Pterodon* (Hyainailourinae). *Pterodon* was a large hypercarnivorous hyaenodont. Teratodontines like *Masrasector* from the Fayum and Taqah were small, agile, fast-moving terrestrial predators (Borths and Seiffert, 2017).

Four early Oligocene species from the Fayum have been recognized as endemic afrotherian insectivores of the tenrecoid group (Seiffert et al., 2007), which includes living golden moles (Chrysochloridae), tenrecs (Tenrecinae), and otter shrews (Potamogalinae). One of them, *Eochrysochloris tribosphenus*, has been identified as an early golden mole (Seiffert

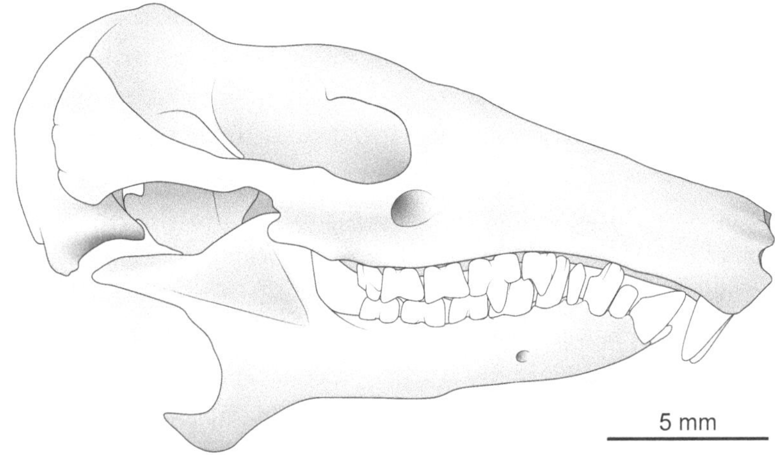

Figure 4.22. The earliest-known golden mole of the chrysochlorid family (afrotherian): *Namachloris arenatans* from the late Oligocene (ca. 27 Ma) of the Eocliff locality, Namibia. Skull of an almost complete skeleton extracted from the limestone matrix after acid preparation by M. Pickford. Drawing by A. Lethiers from Pickford (2015d, figs. 16 and 20).

et al., 2007). However, the Fayum species are poorly known, being represented only by dental remains, and their morphology is distinct, much less specialized than in extant tenrecoids. In particular, these species do not have zalambdodont teeth—that is, the simplified puncturing molars (characterized by a reduction of the metacone of the upper molars and of the talonid of the lower molars) of extant tenrecoids. If the afrotherian insectivores from the Fayum are indeed related to the tenrecoids, they are likely a relict of stem groups rather than members of extant families. The genera *Damarachloris* and *Diamantochloris* from the middle Eocene Namibian site of Black Crow (ca. 43 Ma) have been related to the extant chrysochlorid family (Pickford, 2015c, 2018a, 2019). However, their teeth are much less specialized than chrysochlorids, and they more closely resemble some of the Fayum taxa, such as *Eochrysochloris* and *Widanelfarasia*. The most advanced Paleogene tenrecoids are found in the late Oligocene site of Eocliff (Namibia) **(fig. 4.22)**. The presence of basal groups and the absence of modern families of tenrecoids in the Fayum stands in contrast to the Namibian Eocliff site. It might support a significant difference in age between the sites.

Another group of afrotherian insectivores, the elephant shrews or sengis (macroscelideans), remain discrete in the early Oligocene. These strictly African insectivores are represented in the Rupelian by two small species from the Fayum, known by a few jaw fragments. One of them, *Herodotius pattersoni* (from the Fayum-L41 site), is a member of the basal herodotiine lineage known since the middle Eocene. The other, *Metoldobotes stromeri* (from the Fayum Quarry M), belongs to the poorly known metoldobotine lineage. *Metoldobotes stromeri* has bunodont crushing teeth (i.e., teeth with robust crowns and bulging relief). It is a more specialized macroscelidean that lost the last molar and whose last premolar is large. Discovery of an unpublished skull of *Metoldobotes* from the Fayum has been announced (Seiffert et al., 2007). All these early Oligocene species are stem groups of modern elephant shrews, and there are no known representatives of the extant lineages yet.

The hyracoids (order of living dassies) were at their peak in the early Oligocene. The richest Oligocene sites are by far those of the Fayum of the Jebel Qatrani Formation, particularly the site L41, which alone has yielded nine species. The Libyan site of Jebel al Hasawnah (33 Ma) is also remarkable because it has yielded complete articulated skeletons of early hyraxes **(fig. 4.23; plate 14)**. Such articulated skeletons of mammals are exceptional discoveries from Island Africa. Skeletons of hyraxes from Jebel al Hasawnah belong to young individuals that drowned in a small ancient lake. They all belong to a local population of the species *Saghatherium antiquum* (Gheerbrant et al., 2007), which was originally known by jaw fragments discovered in the Fayum at the beginning of the 20th century. The skeletons found in the Jebel al Hasawnah site show that the Oligocene saghatheriids had a generalized hyracoid morphology and a remarkably similar appearance to extant dassies. They differ mainly in the hands and feet, which retain the first digit and are not unguligrade but digitigrade, which is primitive within the group. The Saghatheriidae are therefore a family of small, generalized terrestrial hyracoids with running adaptations. In total, three families and about 20 species of hyracoids are known during the Rupelian (34–28 Ma). Among them, *Thyrohyrax* is placed at the root of the extant dassie family, the Procaviidae: It is their closest Paleogene stem group.

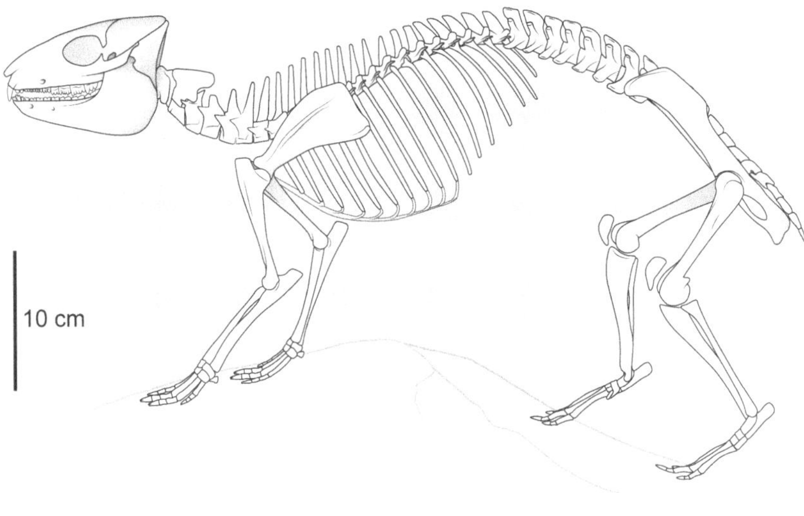

Figure 4.23. Skeletal reconstruction of the early hyracoid *Saghatherium antiquuum* from a specimen found in the early Oligocene of locality Jebel al Hasawnah, Libya **(see also plate 14)**. This early hyracoid had remarkably similar size and appearance to extant dassies, but it was a terrestrial hyracoid adapted to running. Redrawn by A. Lethiers from Gheerbrant et al. (2007).

The Oligocene hyracoids had a wide range of sizes that is unknown today. There were abundant, large tapir-sized hyraxes, such as the geniohyids *Megalohyrax* and *Pachyhyrax*. The titanohyracid, aptly named *Titanohyrax ultimus*, was a giant species that weighed about one metric ton. However, this is a rare species only known by a few teeth. The Oligocene hyracoids colonized a variety of herbivore niches equivalent to those occupied today in Laurasia by both the artiodactyls and perissodactyls. They had remarkably diverse dental specializations and diets. Geniohyids, such as *Geniohyus* and *Bunohyrax*, occupied an ecological niche similar to that of the pig family (the Suidae). They had an elongated snout and were omnivores with bunodont teeth, similar to those of pigs, to eat roots, tubers, and fruits. *Saghatherium*, *Thyrohyrax*, and *Megalohyrax* had teeth that were bunodont, lophodont, and selenodont, indicating a varied diet as in the living dassies. They fed on grasses, leaves, fruits, twigs, and bark. *Megalohyrax* was a generalist Oligocene hyracoid very well adapted

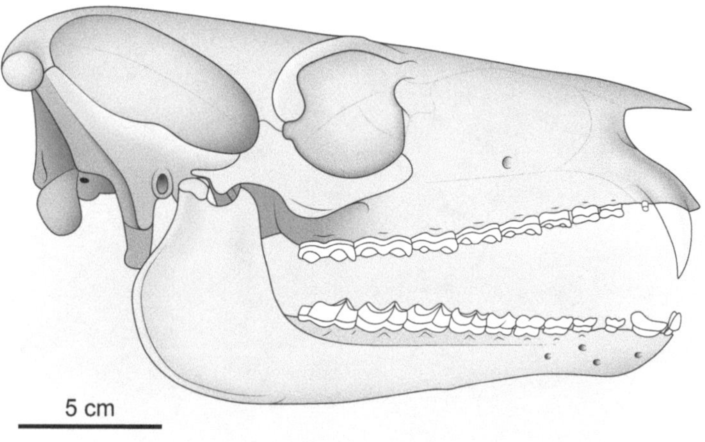

Figure 4.24. Skull of the antelope-like hyracoid *Antilohyrax pectidens* from the Fayum-L41 locality, 34 Ma, Egypt. This gazelle-sized hyracoid (body mass 35 kg) displays a remarkable resemblance in its teeth and limbs to bovid artiodactyls (Laurasiatheria), such as extant antelopes, and it is one of the best examples of convergent evolution among African and Eurasian ungulates. This early hyrax was a cursorial folivorous browser. Skull redrawn by A. Lethiers from De Blieux and Simons (2002).

ecologically to African environments. It evolved over a long period until the late Oligocene (Chilga site in Ethiopia) without much morphological change compared to other hyracoids. Titanohyracids, such as *Titanohyrax*, *Selenohyrax*, and *Antilohyrax* (**fig. 4.24**), had selenodont teeth with crescent-shaped crests and tubercles adapted to a folivorous diet, as in ruminant artiodactyls (e.g., bovids). *Antilohyrax* was specialized for running in leaps like today's springbok (jumping antelopes). These Oligocene hyracoids show varying degrees of sexual dimorphism. *Thyrohyrax*, *Megalohyrax*, *Pachyhyrax*, and *Bunohyrax* have a peculiar swollen mandible with a hollow internal chamber with an open window in the bone. In *Thyrohyrax*, the mandibular window is open only in the males. The mandibular chamber served as a sounding board ("vocal resonator" in De Blieux et al., 2006) to amplify the calls of these hyracoids that would have been echoed in the ancient forests of the Island Africa.

The most iconic and amazing Oligocene mammal of Island Africa is without doubt the embrithopod *Arsinoitherium zitteli* (**fig. 4.25**). The ge-

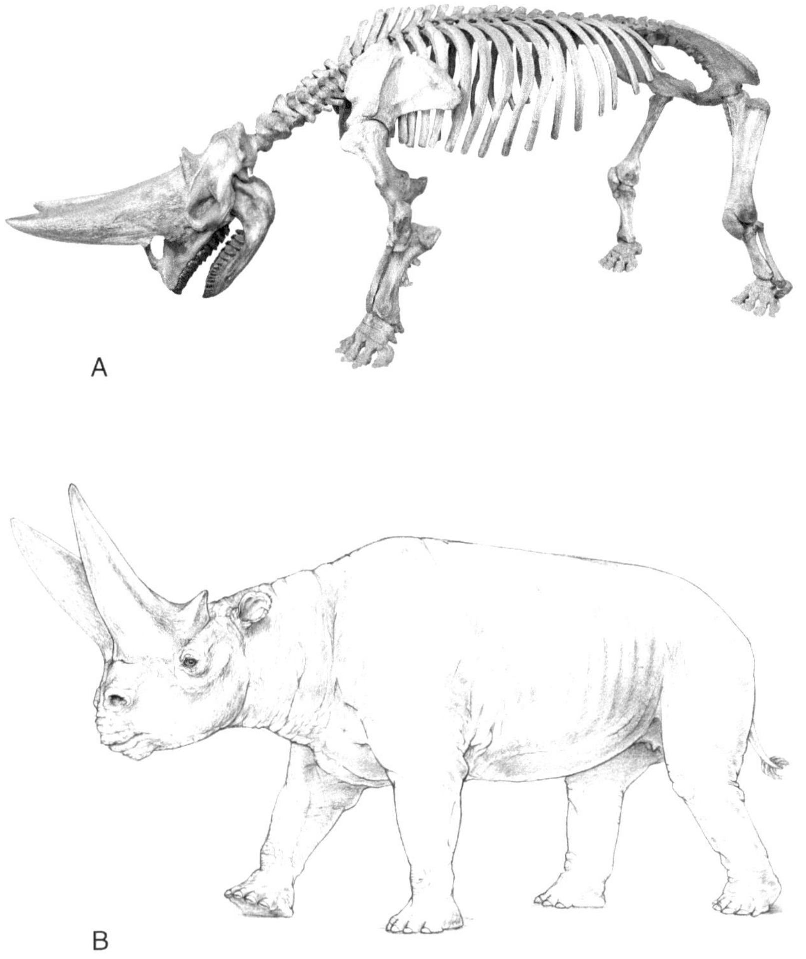

Figure 4.25. The embrithopod *Arsinoitherium zitteli* from the early Oligocene of the Fayum, Egypt, 34 Ma, was one of the prehistoric Big Five of Island Africa. **A.** Skeleton reconstructed and reassembled (length about 3 m; height at shoulder 1.75 m). **B.** Life restoration of *Arsinoitherium zitteli* from the Fayum, Egypt, 34 Ma. Skeletal reconstructed by C. W. Andrews and exhibited in the Natural History Museum in London; photo by DagdaMor, https://commons.wikimedia.org/wiki/File:Arsinoitherium_MNH_2022.png (Wikimedia Commons, CC BY-SA 4.0) (**A**). Drawing by C. Letenneur (**B**).

nus *Arsinoitherium* was already present in the late Eocene, but we know it from a poorly known and unidentified species. The most abundant and complete fossils of *Arsinoitherium zitteli* were found in the early 20th century in the early Oligocene Fayum sites. *Arsinoitherium zitteli* has been found in all levels, from the base to the top, of the Jebel Qatrani Formation, which extends in time from 34 to 29 Ma. It is a relatively frequent species. In fact, the fossil remains of *Arsinoitherium zitteli* found in the Fayum are among the most complete ever discovered in the African island for any mammal. This allowed the paleontologist C. W. Andrews of the British Museum (London) to study and reconstruct the entire skeleton of the animal (Andrews, 1906), an exceptional case for mammals from Island Africa. *Arsinoitherium* has also been found in the early Oligocene of Oman, Angola, Tunisia, and Libya (Dur at Talah). A second species, *Arsinoitherium andrewsi*, has been described from the Oligocene of Fayum, but it is generally considered to correspond to large individuals of *Arsinoitherium zitteli*. *Arsinoitherium* is emblematic of the herbivorous Oligocene megafauna of Island Africa, together with early elephants (barytheres and palaeomastodonts) and large early hyracoids. It was a spectacular animal, both by its size (shoulder height of 1.8 m—close to the rhinoceros) and by its strange look, with a pair of enormous nasal horns (flanked by a small posterior pair on the frontals). Its body size and nasal horns were variable among individuals, in relation to strong sexual dimorphism that is common in gregarious mammals. Like the elephants, *Arsinoitherium* had a graviportal skeleton. This is a specialization of the large animals in which the limbs are erected into a vertical column and placed under the body to carry it. *Arsinoitherium* was a heavy grazer, not well adapted to running. In fact, "embrithopod" means "heavy foot." It was previously assumed that *Arsinoitherium* was a semiaquatic animal living in swamps, but the analysis of the isotopic composition (carbon and oxygen) of the teeth indicates this was a terrestrial mammal. *Arsinoitherium* has highly specialized teeth of a folivorous (leaf-eating) animal; the teeth are lophodont with sharp transverse crests and hypsodont with a high crown. Its lophodont dental morphology is convergent with that of hyracoids and proboscideans: The similar shape is constructed in a different, nonhomologous way in embrithopods, being derived from the peculiar hyperdilambdodont pattern. Premolar and molar teeth display very different morphology in the tooth row of *Arsinoitherium*. This is related to a particular chewing function in *Arsinoitherium* in which the food was crushed more finely by the premolars than by the molars, in contrast to most mammals (Court, 1992). The very deep palate of *Arsinoitherium* indicates that its tongue was large and played an important role both in seizing the plants it ate and in helping to chew them. It is thought that *Arsinoitherium* was a very selective herbivore. It preferred large fruits, the fossils of which abound in the Fayum sites. Besides these remarkable specializations, *Arsinoitherium* preserves primitive characters lost in other African ungulates. It has, for example, a complete dental formula with the presence of the canine and four premolars per half jaw; this is very different from, for example, Oligocene elephants that lived at the same time and had lost the canines and anterior premolars.

In the early Oligocene, the elephant order (proboscideans) is represented by moeritheres and the first species of the elephantiform group that includes modern elephant-like proboscideans. The stem proboscideans that have lophodont cutting teeth, such as numidotheriids and barytheriids, were in decline. Their last-known species is *Omanitherium dhofarensis* from Oman (Seiffert et al., 2012). This species is still poorly known, being represented mostly by jaws and teeth. It has been related to the origin of the deinotheres, but new findings suggest that it is more likely to be a barythere close either to *Barytherium* (Tassy, 2015) or to *Arcanotherium* from the late Eocene of Libya (Al-Kindi et al., 2017). The latest moeritheres are from the early Oligocene. They are represented by *Moeritherium trigodon* from the Fayum. Like *Moeritherium lyonsi* from the late Eocene, *M. trigodon* was an amphibious animal that fed on aquatic plants.

The early Oligocene is a key period in the evolution of proboscideans, with the appearance of the first proboscideans of modern elephantine aspect, the elephantiforms, represented by the early palaeomastodontid family. Two palaeomastodont species, *Palaeomastodon beadnelli* and *Phiomia serridens* **(fig. 4.26)**, were identified in the Fayum (notably in Quarries A–B) as early as the beginning of the 20th century. They differ from each other in the morphology of their teeth, cutting (lophodont) in *Palaeomastodon* and crushing (bunodont) in *Phiomia*. *Phiomia* is also more advanced in its longer mandibular symphysis. Recent discoveries of more fragmentary fossils of *Palaeomastodon* and *Phiomia* were also made in Libya, Oman, Egypt, and Tunisia. *Palaeomastodon* and

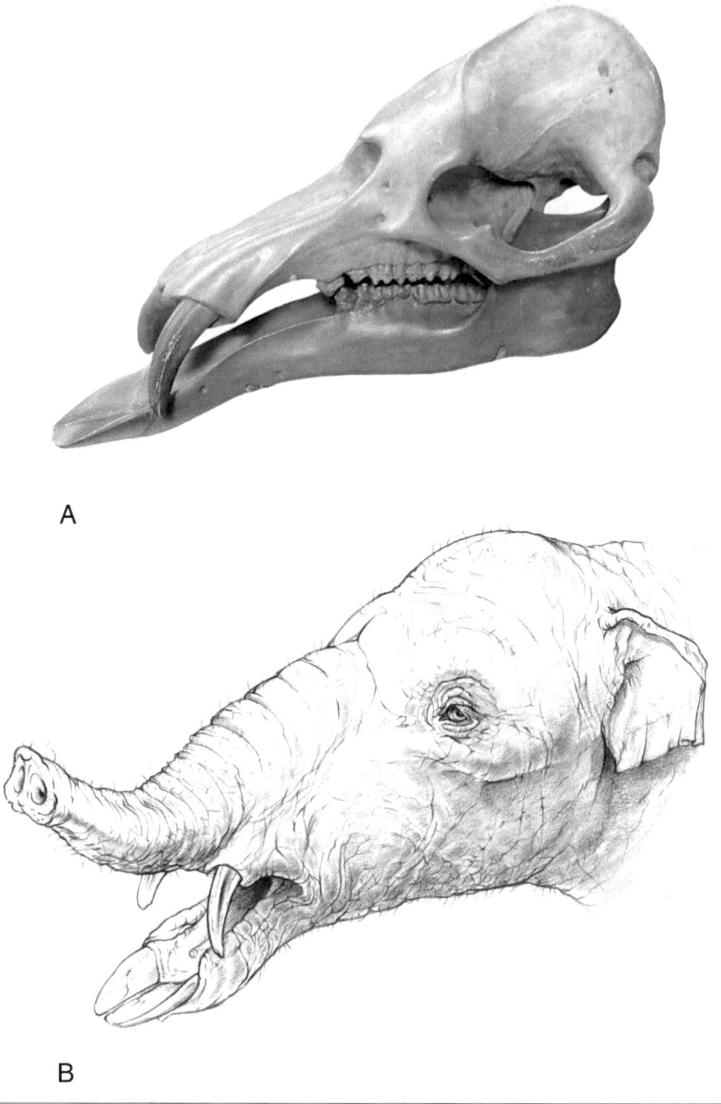

Figure 4.26. *Phiomia serridens*, from the early Oligocene (34 Ma) of the Fayum (Egypt), is the first-known elephantiform (Proboscidea), with its coeval *Palaeomastodon*; both *Phiomia* and *Palaeomastodon* are related to the origin of modern proboscideans of elephantine appearance (collectively called elephantiforms). **A.** Skull of *Phiomia serridens* (length with tusks 90 cm). **B.** Restoration of the head of *Phiomia serridens*. Reconstruction (model) exhibited in the paleontology gallery at the National Museum of Natural History in Paris, donated by C. W. Andrews in 1908; photo E. Gheerbrant (**A**). Drawing by C. Letenneur (**B**).

Phiomia are placed at the root of the great evolutionary radiation of the modern elephantoid proboscideans that occurred during the Miocene. They are two successive stem groups of all the elephantiforms. *Palaeomastodon* (shoulder height of 2 m; body mass of 2 to 2.5 metric tons) reached the size of the modern Asian elephant, and *Phiomia* was a bit smaller (1.5 m tall at the shoulder). These two palaeomastodonts retain primitive traits lost in later elephantiforms, such as a long and low skull and teeth that were replaced vertically. However, they also had important advanced characters that announce the modern elephants, such as teeth with three well-formed lophs (transverse crests that will multiply in modern elephants), a skull raised at the back (occiput), nostrils retracted back toward the orbits indicating the presence of a small trunk, two pairs of tusks that begin to elongate, and a robust graviportal skeleton, in which the body is carried by limbs raised into vertical columns as in extant elephants. Analyses of the isotope composition of the teeth indicate that *Phiomia*, like *Moeritherium*, ate mangrove plants (Clementz et al., 2008).

The sea cows (sirenians) from the early Oligocene of Africa remain poorly known. Only one species from the Fayum, *Eosiren imenti*, has been identified (Domning et al., 1994). *Eosiren* is a "halitheriine"—that is, a member of the closest stem group of extant dugongs and manatees. It is known in Africa since the middle Eocene. If the known fossil record is representative, the prorastomids disappeared from Africa before the Oligocene.

During the Paleogene, it was in the early Oligocene that the rodents were the most diversified in Island Africa. There are about 25 described species belonging to about 10 families. The classical sites of the Fayum (Egypt) have yielded the best-preserved fossils. Recent important discoveries also have been made in Morocco (Dakhla), Libya (Zellah), and Oman (Taqah), but these are more fragmentary fossils comprising mainly isolated teeth.

The African evolutionary radiation of the anomaluroids, from which the extant flying squirrels (anomalurids) evolved, was well underway during the Oligocene with the presence of at least four species in two families in the Rupelian sites. An intriguing point is that the Oligocene anomaluroids present in Oman, Libya, and Morocco are absent from the Fayum sites. This may be because they lived in more closed forests (Heritage et al., 2016). Most stem anomaluroids (zegdoumyids and nementchamyids), with the only exception of *Dakhlamys ultimus* from Dakhla, disappeared before the Oligocene. They are early evolving lineages predating the non-anomalurids and the extant families of zenkerellids and anomalurids (**fig. 4.27**). The Miocene genus *Nonanomalurus* has been found in the early Oligocene Dakhla site (Morocco, 34 Ma) (Marivaux et al., "Anomaluroid Rodents," 2017). It is the closest stem group of the two extant zenkerellid and anomalurid families. The earliest flightless scaly-tailed squirrels of the zenkerellid family have been found in the early Oligocene (34 Ma) of Dakhla (Morocco) and Zellah (Libya) with *Oromys* and *Prozenkerella*, respectively. *Paranomalurus*, also known from the Miocene, and *Argouburus* from Dakhla are early Oligocene stem anomalurid genera. They indicate that the anomalurid family specialized for gliding existed for at least 34 million years and has been very conservative (e.g., *Paranomalurus* lineage spanning for over 10 million years). Furthermore, *Paranomalurus* from Dakhla is more closely related to the extant genus *Anomalurus* (the large flying squirrels) than to *Idiurus* (the flying mouse). This argues that the diversification of extant anomalure genera dates to before 34 Ma, as early as the Eocene.

The African hystricognath rodents (i.e., the phiomorphs) are even more diversified in the Oligocene. They are represented by about 20 species. The first discoveries were made in the Fayum sites at the beginning of the 20th century, and since then they have been found in all the main Rupelian African sites yielding small mammals. Some groups are new in the Oligocene (Gaudeamuridae and Metaphiomyinae). Apart from the Gaudeamuridae, all Oligocene phiomorphs are successive stem groups of the thryonomyoids. *Birkamys* and *Mubhammys* from the Fayum-L41 site have lost the last premolar, which is a key character of the extant African phiomorphs (i.e., the thryonomyids, petromurids, and bathyergids). *Metaphiomys* and in particular *Monamys* from the Fayum (Quarry I, 30–29 Ma) are the closest Paleogene relatives of the two

Figure 4.27. The extant African "flying squirrel" *Anomalurus* of the endemic African family Anomaluridae (not to be confused with North American flying squirrels, which are a separate group of the northern sciurids). Anomalures are specialized arboreal and nocturnal African rodents, and *Anomalurus* is the most specialized, moving by gliding between trees to forage for plants (fruits, flowers, leaves). Drawing by C. Letenneur.

extant families of cane rats (thryonomyids) and dassie-rats (petro-murids). The gaudeamurids that appear in the Fayum-L41 site have divergent specializations from extant African phiomorphs: They are the closest African relatives of the South American caviomorph rodents.

Finally, the muroid rodents of the hamster family—the Cricetidae—appeared in Africa in the early Oligocene. They have been found only in the locality of Taqah (Oman). They testify to a dispersal from Eurasia, where the group is known since the Eocene. The presence of Oligocene cricetids in Oman, and only there in Africa, probably reflects a geographic particularity: It is likely explained by an early localized connection of the Arabian Peninsula with Asia (i.e., via the Iranian region) **(see fig. 6.3)**.

During the Oligocene, the Arabo-African primates rapidly diversified and flourished. They are the most abundant mammals in the Rupelian sites of Island Africa, along with the rodents and hyracoids. The predominant group, discovered in the Fayum at the beginning of the 20th century, is that of the anthropoid primates. Most of their fossils come from the Fayum and Oman, and it is in the Fayum that the most complete remains (skulls and postcranial skeletal bones) have been found.

Strepsirrhines that lived in Island Africa have been found fairly recently, with species of extinct adapids and extant lorisiforms. Their first fossils were discovered in 1993 in the early Oligocene Omani site of Taqah (Gheerbrant et al., 1993). The early Oligocene species *Aframonius dieides* of the extinct strepsirrhine adapid family was found in 1995 in the Fayum-L41 site (Simons et al., 1995). It was a relatively large adapid, weighing 1.5 kilograms, and it was a specialized folivore. It shows notable convergences with the anthropoids—for example, in the presence of a large and sexually dimorphic canine, a fused mandibular symphysis, upper molars with a strong hypocone (the fourth molar tubercle with a crushing-grinding function), and a second premolar in the process of reduction. *Aframonius* belongs to the caenopithecine adapid lineage. This lineage is of European origin, and it is recorded in Africa from the middle and late Eocene with *Namadapis* and *Masradapis*.

Modern strepsirrhines from the early Oligocene sites of Africa are found in Taqah (Oman) with a few teeth of the two species *Omanodon minor* and *Shizarodon dhofarensis*, and in the Fayum-L41 site (Egypt)

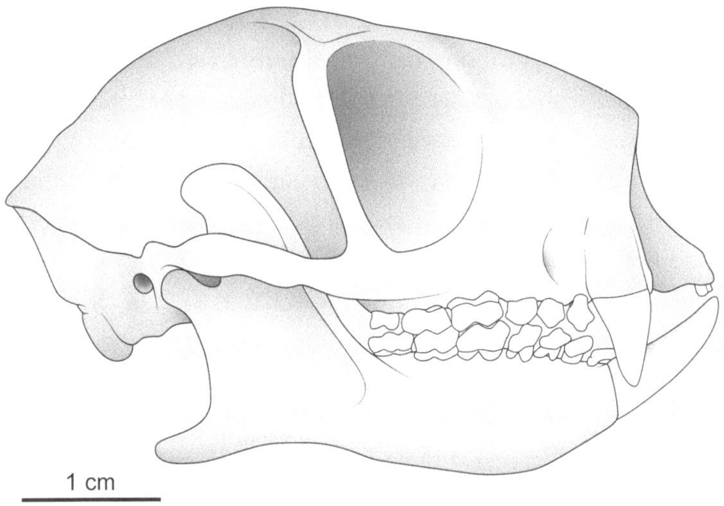

Figure 4.28. *Plesiopithecus teras*, from the early Oligocene of the Fayum-L41 (Egypt, 34 Ma), is the oldest relative of the aye-aye lemurs (the Chiromyiformes) of Madagascar. *Plesiopithecus* represents the best-known fossil evidence of the African origin of the Malagasy aye-ayes. Skull redrawn by A. Lethiers from Simons and Rasmussen (1994).

with "*Anchomomys*" *milleri* (genus identification questionable), which is better known. These species are either basal strepsirrhines that lack a tooth comb in the lower jaw or early lorisiforms. Although their incisors are unknown, their relationships to modern strepsirrhines are indicated by the morphology of the lower premolars and canines, which are elongated anteriorly as in the tooth comb of the extant strepsirrhines species. *Wadilemur elegans* from the Fayum-L41 site (latest Eocene or earliest Oligocene, ca. 34 Ma) is a typical galagid lorisiform.

Another important and unexpected discovery was made in 1992 in the Fayum-L41 site: The remains of the skull and jaws of *Plesiopithecus teras* **(fig. 4.28)**. This species exhibits specialized characters that are known only in the living Malagasy aye-ayes (genus *Daubentonia* of the Chiromyiformes) (Godinot, 2006; Gunnell et al., 2018). Aye-ayes are peculiar nocturnal lemurs that feed on xylophagous insect larvae present under the bark of trees. Aye-ayes catch larvae in the wood with specialized, elongated, slender, and clawed fingers. *Plesiopithecus* shares with the aye-ayes a rodent-like mandibular morphology that bears a very large, anteriorly inclined incisor (which is associated with a reduction of the other incisors and canines). The homology of the enlarged anterior tooth of *Plesiopithecus*, either a canine or an incisor, has long remained uncertain. However, a small tooth-comb-like canine has recently been shown to be present behind this tooth (Gunnell et al., 2018). *Plesiopithecus* from the Fayum-L41 site and *Propotto* from the Miocene of East Africa are the only known representatives of the two Malagasy groups of lemurs and aye-ayes in Africa, and *Plesiopithecus* is their earliest-known representative.

The poorly known afrotarsiid family was first discovered with a mandible of *Afrotarsius chatrathi* in the early Oligocene Fayum Quarries M–P (Simons and Bown, 1985). Additional postcranial fossils of this species were reported from the same sites by Rasmussen and colleagues (1998). *Afrotarsius* is also known from the late Eocene of Libya. *Afrotarsius*, as its name indicates, was initially considered to be related to the Asian group of the tarsiers (tarsiifoms) (Rasmussen et al., 1998; Seiffert, 2012; Seiffert et al., 2018), but other studies support them being related to the stem anthropoid family Eosimiidae (Beard, 1998; Jaeger et al., "Late Middle Eocene," 2010). The two hypotheses agree that afrotarsiids originated in Asia and reached the African island in the late Eocene.

After a first timid appearance in the Eocene with the successive genera *Amamria* and *Biretia*, the anthropoid primates exploded in diversity and abundance in the Oligocene in Africa. Nearly 20 species in four families are known from about 10 localities, mostly in North Africa and Oman. These species belong to three distinct groups of anthropoids: (1) the parapithecids (parapithecoids), which are the stem group of South American (Platyrrhini) monkeys and Old World monkeys and apes (Catarrhini); (2) the proteopithecids, which are a sister group to the parapithecids; and (3) the stem catarrhine monkeys of the families Oligopithecidae and Propliopithecidae. These anthropoids share many skeletal features, particularly in the dentition. For example, the molars are bunodont (crown with bulging tubercles used for crushing nuts), the incisors are spatulate and vertical, and the first lower premolar is large and has a sharp anterior wear facet sharpened by the upper canine.

In the skull the orbit is closed at the back, and the two bones of the frontal and mandible (dentaries) are fused together. Fossils also evidence the presence of sexual dimorphism in these anthropoids (i.e., a variation in the development of certain skeletal characters related to sex of the individuals). For example, the canines are much stronger in males than in females.

Proteopithecids are represented by the species *Proteopithecus sylviae* and *Serapia eocaena* from the Fayum-L41 site. *Proteopithecus sylviae* is known by numerous well-preserved fossils, including skulls and limb bones. This marmoset-sized species resembles the New World monkeys (Platyrrhini, the monkeys with prehensile tails) in their morphology. *Proteopithecus* was a small frugivorous diurnal primate that lived in colonies in trees, where it moved rapidly by running and jumping from branch to branch. It shares a common ancestor with the parapithecids.

Oligocene parapithecids were very diverse in Island Africa. They include five genera in two subfamilies (parapithecines and qatraniines). Numerous fossils have been found in the Fayum. They have helped us understand the parapithecids' locomotion and feeding specializations, and their relationships. *Abuqatrania* and *Qatrania* are two small basal parapithecids from the lower levels of the Jebel Qatrani Formation (34–33 Ma). *Apidium* and *Parapithecus*, from the upper levels of the Jebel Qatrani Formation, are more recent (30–29 Ma) and are more evolved. *Apidium* is the most abundant and best-known primate from the Fayum sites. The parapithecids were small diurnal and arboreal anthropoids about the size of a *Saimiri* (body mass up to 3 kg in *Parapithecus grangeri*). They moved in groups by agile jumps from branch to branch in the tropical forests of Island Africa. In anatomy and lifestyle, they resembled the small living New World monkeys, such as the squirrel monkey *Saimiri*.

The third major group of primates known to have lived in Island Africa during the Oligocene comprise the catarrhines, which include the Old World monkeys (cercopithecoids) and the apes (hominoids: lesser apes, great apes, and humans) **(see figs. 4.29, 4.30, 5.12; plate 15)**. In the Rupelian there are two stem catarrhines families: the oligopithecids and the propliopithecids.

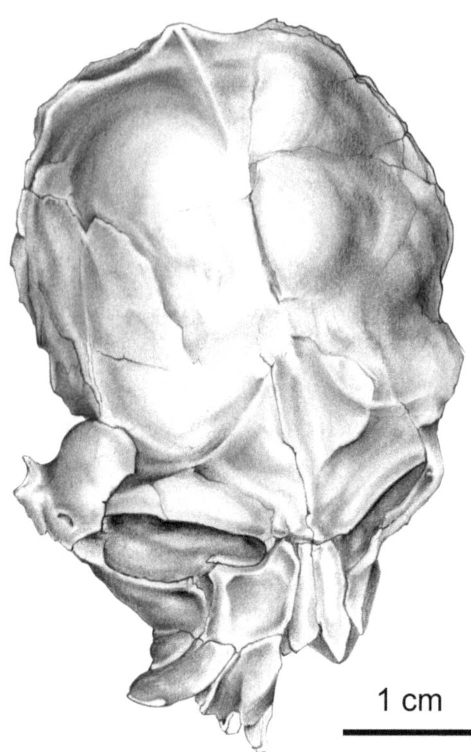

Figure 4.29. Skull, seen from above, of the anthropoid primate *Catopithecus browni*, discovered in the Fayum-L41 site (Egypt, 34 Ma). *Catopithecus* is the oldest and basal-most known catarrhine, the group of Old World monkeys and apes. Redrawn by C. Letenneur from Simons and Rasmussen (1996, fig. 2).

Oligopithecids are not as well known as propliopithecids. Their first discovery was made in 1962 by E. L. Simons in the Quarry E of Fayum with a broken lower jaw of *Oligopithecus savagei* (Simons, 1962). This rare species remained enigmatic for a long time, until new discoveries 30 years later in the Fayum and in Oman. The most remarkable discovery regarding the oligopithecids was that of abundant fossils, including skulls and postcranial bones, of *Catopithecus browni* (**fig. 4.29**) from the Fayum-L41 site in the lowermost part of the Jebel Qatrani Formation (Simons, 1989). *Catopithecus* is a small tamarin-sized species (body mass 400–800 g). It is the oldest-known catarrhine, possibly along with the poorly known genus *Talahpithecus* from Libya, of uncertain age. Another species of *Oligopithecus, O. rogeri,* was found in the Omani site of Taqah, where it is the most abundant primate (Gheerbrant et al., 1995). Finally, a poorly known "dwarf" oligopithecid species (body mass about 200 g) was reported in 2013 from Quarry M, 30–29 Ma, from the upper sequence of the Jebel Qatrani Formation (Seiffert and Simons, 2013). This is the last-known oligopithecid; the family went extinct soon afterward, before the late Oligocene.

Propliopithecids have been called the "dawn apes" (Simons, 1968). Three genera, *Propliopithecus*, *Moeripithecus*, and *Aegyptopithecus*, have been discovered. In the Fayum they appear later than the oligopithecids, in the middle part of the Jebel Qatrani Formation (Quarries G–V), around 31 Ma, with *Propliopithecus ankeli*. However, older propliopithecids, such as "*Moeripithecus markgrafi*" (questionable genus and species attribution) **(plate 15)**, were discovered by Thomas and colleagues (1991) in the Taqah site (Oman), where they are associated with oligopithecids. The association of propliopithecids and oligopithecids that occurred in the Taqah fauna is only recently known in the Fayum, and only in the upper levels of the Jebel Qatrani Formation. The Taqah propliopithecids are interesting because they are more primitive in several respects than those known from the Fayum. The most famous propliopithecid is *Aegyptopithecus zeuxis*. It was found in 1965 by E. L. Simons in the upper Jebel Qatrani Formation (Quarries I and M, 30–29 Ma) from the Fayum, with very nice fossils including skulls and limb bones (Simons, 1965). *Aegyptopithecus* was a slow, arboreal, quadrupedal early catarrhine that resembled the present-day South American howler monkey (*Alouatta*), but it was more robust **(fig. 4.30)**. *Propliopithecus* had a prehensile foot with an opposable thumb for holding onto branches, and it moved through the trees by hanging from the branches.

The Oligocene bats of Africa were highly diverse. There are nearly 10 species in seven families (Philisidae, Emballonuridae, Vespertilionidae, Hipposideridae, Nycteridae, Myzopodidae, and Megadermatidae). All are microbats (microchiropterans). However, their fossils are rare in the Oligocene mammal sites, and they have been found only in the Fayum (Egypt) and in Taqah (Oman). All Oligocene families are known since the Eocene, although with a gap in the late Eocene. They are all modern

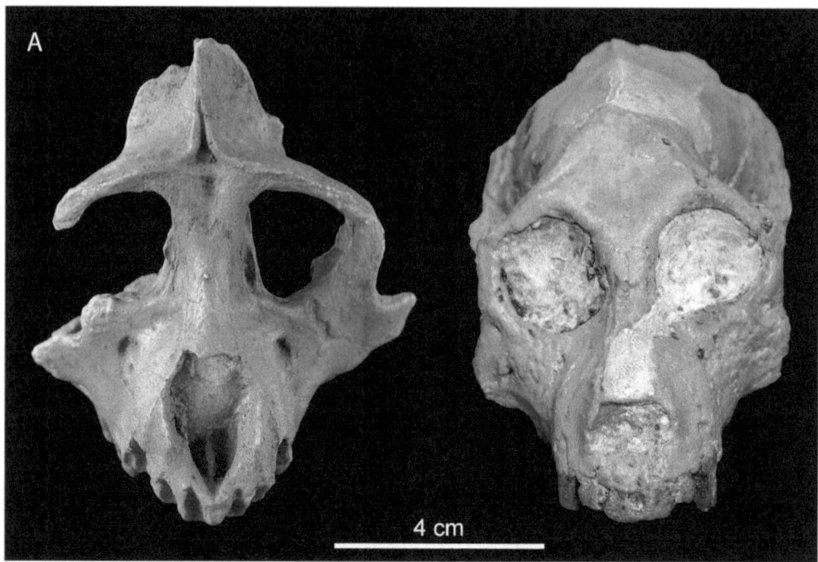

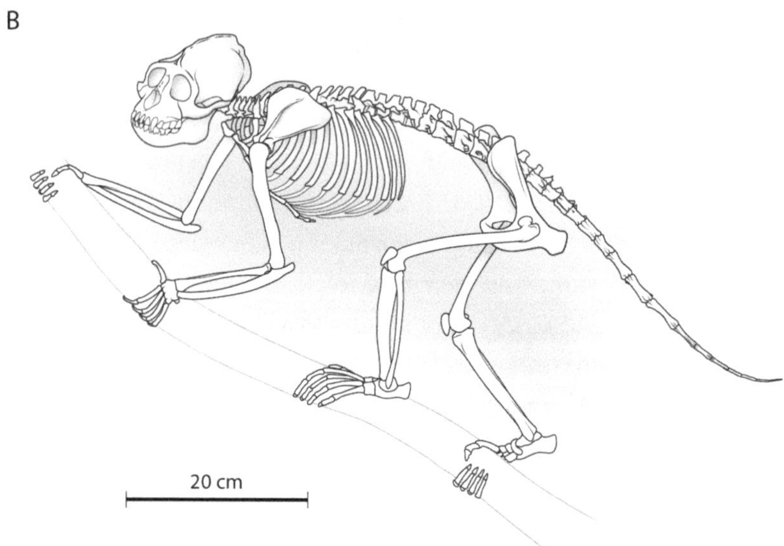

Figure 4.30. The most famous stem catarrhine primate, *Aegyptopithecus zeuxis*, discovered in the Fayum, Egypt by E. L. Simons. *Aegyptopithecus* was a large ancestral catarrhine monkey, with a body mass estimated to be four to seven kilograms. It evolved in the early Oligocene tropical forests of Island Africa. **A.** Skull of *Aegyptopithecus zeuxis* (*left*, male; *right*, female) from the Fayum early Oligocene beds of about 30 Ma. *Aegyptopithecus* is an early fossil catarrhine resembling South American monkeys and showing strong sexual dimorphism—that is, males and females showing differences in size and morphology. The difference is most obvious here in the strong sagittal crest at the top of the skull in the male. **B.** Reconstruction of the skeleton of *Aegyptopithecus zeuxis*. Photos of casts by L. Cazes (**A**). Redrawn by A. Lethiers from Fleagle (2013) (**B**).

families, except for the extinct Philisidae. These Oligocene microbats are insectivorous or carnivorous. *Vampyravus orientalis* is the first discovered bat from Island Africa. It was identified in 1910 by the German paleontologist Max Schlosser from a single humerus found in the Fayum. It is the size of the extant fruit bats (estimated body mass 120 g), and it is the largest microchiropteran known in the Fayum. The exact relationship of *Vampyravus* is uncertain, but it belongs to a stem group of a modern family, perhaps the emballonurids (sheath-tailed bats) or the rhinopomatids (mouse-tailed bats), the former being already known in the Eocene. *Saharaderma pseudovampyrus* from the Fayum-L41 site is the oldest-known species of the "false vampire" family (Megadermatidae). The Madagascar family of "sucker-footed bats" (Myzopodidae) is known by an Oligocene African species of the genus *Phasmatonycteris*, which is recorded from earlier late Eocene levels of the Fayum (Gunnell et al., 2014).

Scaly anteaters, or pangolins (pholidotes), made a brief African appearance in the upper Jebel Qatrani Formation in the Fayum. They are very poorly known, and their identification in the Fayum needs to be verified. Two phalanges found in Quarries M and L-12, dated at 30–29 Ma, are attributed by Gebo and Rasmussen (1985) to an unidentified species of the extant manid family of the pangolins. In support of this identification, these paleontologists report the presence of burrowing tracks similar to those of extant pangolins in fossilized termite mounds present in the same levels of the Jebel Qatrani Formation where the two phalanges were found. The family Manidae has a predominantly Laurasian distribution in the Paleogene, and its oldest records are from the middle Eocene (45 Ma) of Europe based on the exceptional *Eomanis* specimens from the Messel locality (Germany). The presence of manid pangolins at the Fayum sites, if confirmed, is the result of a dispersal into Africa from the Laurasian areas, and possibly from Europe, prior to the early Oligocene.

The African Paleogene anthracotheres are overall not very well known; their fossils are few and mostly fragmentary. However, a notable exception is the Fayum sites, where they are among the most abundant large herbivorous mammals. More than 2,000 specimens have been found throughout the entire fossiliferous series of the Jebel Qatrani Formation

(Sallam et al., 2016), including skulls, jaws, and even partial skeletons. Anthracotheres were semiaquatic grazing euungulates ("true ungulates") of Asian origin. In the early Oligocene of Africa they diversified with the appearance of seven species in two new genera, *Qatraniodon* and *Nabotherium* (formerly named *Rhagatherium*). *Nabotherium* represents an anthracothere lineage with bunodont crushing teeth. The genera *Bothriogenys* and *Qatraniodon* are folivorous anthracotheres with sharper teeth bearing inflated and crescent-like cusps ("bunoselenodont" teeth). *Qatraniodon* is the smallest and rarest anthracothere from the Fayum. It is known by a single specimen. *Nabotherium*, with its shorter rostrum, large projecting canines, simple premolars, and low bunodont molars, was probably a more eclectic frugivorous and herbivorous anthracothere than the semiaquatic *Bothriogenys*. The species *Bothriogenys gorringei* from the Fayum is known in Eoridge (Namibia), which is more consistent with the early Oligocene age of this site.

The "toothed whales," or archaeocetes, that were abundant in the Eocene in Africa disappeared in the Oligocene.

Late Oligocene, Chattian, 28–23 Ma

This period has been brought to light by a series of crucial new paleontological discoveries made only since the beginning of our third millennium, especially in East Africa. Today, we know of a dozen fossil localities that are spread throughout the Chattian. The earliest sites are in Ethiopia (Chilga, 27 Ma) and Eritrea (Dogali, 27 Ma), and the most recent are in Kenya (Losodok, 25–24 Ma). The Namibian site of Eocliff, initially considered late Eocene, might instead also be late Oligocene. Rodents such as *Metaphiomys* and *Neophiomys* indicate an age of Eocliff later than the Eocene, and according to Bronner and colleagues (2024, fig. 13), the stem chrysochlorid *Namachloris* indicates an age close to 27 Ma (i.e., late Oligocene). Some of the African late Oligocene sites are well dated by absolute ages obtained from radiometric studies, particularly for localities from the East African volcanic regions (Chilga, 27 Ma; Dogali, 27 Ma; Losodok, 25–24 Ma; Mai Gobro, 23.5 Ma). Thirty-three identified species in eight orders of placental mammals are known in the late Oligocene sites from Africa. The most diverse late Oligocene African groups are the hyracoids, elephants (proboscideans), and rodents. There is also a diversity of African hyaenodonts.

The late Oligocene is a period of evolutionary continuity in Africa. It was the last period of endemic evolution in insular Africa. Most of the previously known early Oligocene lineages are present, and several, such as elephant relatives (proboscideans) or catarrhine primates, further diversified. Some endemic groups, such as proboscideans and embrithopods, evolved into huge species. The late Oligocene was also a period of evolutionary transition between the primitive faunas of the Paleogene and the modern ones of the Neogene. The period is marked by the first emergence of some modern groups that flourished during the Neogene. They originated from local, endemic evolution of older African stem groups. The most important modern lineages emerging in Africa during the late Oligocene include hyraxes such as pliohyracids, the elephantoids (also called elephantimorphs) such as true mastodons and gomphotheres (the stem group of the extant elephant family), and catarrhine monkeys (cercopithecoids) and apes (hominoids). Virtually no new mammals of Laurasian origin are known from this period. This is surprising, especially for the end of the period (24–23 Ma), when the Arabo-African island was very close to Eurasia. It is also surprising since we know that there are older Rupelian fossils of Asian taxa found in Arabia, such as the cricetid rodents from Taqah (Oman), that testify to this intercontinental proximity. It is likely only a matter of time before we find fossils of newcomer Asian mammals in the late Oligocene of Arabo-Africa.

Indeed, the late Oligocene is a period of conspicuous fossil gaps. Among the mammals already known in the preceding Rupelian period, and for most known later in the Miocene, there are no marsupials, no ptolemaiids, almost no sea cows (sirenians), no bats, no rodents of the endemic group of flying squirrels (anomaluroids), and no cetaceans (whale group). Some of these gaps in the fossil record have been partially filled by new discoveries at the Eocliff site, in Namibia, which may be late Oligocene. They include especially numerous fossils of macroscelideans, tenrecoids, and rodents.

The hyaenodont carnivorous placentals from the late Oligocene of Africa are not as well known as those from the Rupelian. They are represented by four species from three old Oligocene subfamilies. *Pakakali*, found in Tanzania (Rukwa site, 25 Ma), is a small hyaenodont (body mass <10 kg) that hunted small prey, vertebrates and invertebrates. *Mlanyama*, from Losodok (Kenya, 25–24 Ma), is a highly specialized hypercarnivorous predator of medium size. Its robust lower jaw, fused in the front, was capable of very powerful bites. Hyaenodonts survived late in Africa, across the Oligocene–Miocene boundary.

Most known late Oligocene afrotherian insectivores come from the Namibian Eocliff site, which has yielded abundant and well-preserved fossils. The diversity of the Eocliff afrotherian insectivores is remarkable. They include the forebears of the extant families of golden moles (Chrysochloridae), with the stem genus *Namachloris*, and of otter shrews (potamogalines) and tenrecs (tenrecines), with the stem genera *Namagale*, *Sperrgale*, and *Arenagale*. *Namachloris arenatans* is particularly noteworthy because it is known by extraordinarily well-preserved fossils, including several skulls and a large part of the postcranial skeleton showing specialized traits that characterize the golden mole family Chrysochloridae (Mason et al., 2018). *Namachloris* is thus the earliest-known chrysochlorid. Similarly, fossils of elephant shrews (macroscelideans) are also plentiful at Eocliff, with over 1,000 cranio-dental and postcranial elements. They illustrate a high diversity of elephant shrews with the presence four genera and six species (Senut and Pickford, 2021). They include stem macroscelidids, myohyracines (extinct Miocene macroscelidids), and the earliest relatives of the extant subfamily Rhynchocyoninae (Senut and Pickford, 2021). The locality of Rukwa, Tanzania, also yielded elephant shrews, although their fossils are scarce and fragmentary (Stevens et al., 2022). As in the Eocliff site, they include the earliest myohyracines (*Rukwasengi*) and rhynchocyonines (*Oligorhynchocyon*).

The late Oligocene hyracoids are represented mostly (eight species) by the three ancient Paleogene families (Geniohyidae, Titanohyracidae, and Saghatheriidae) and by two species of a new family that flourished in the Neogene, the pliohyracids. The Paleogene families are represented by early genera already known in the Rupelian of the Fayum. One exception is the late geniohyid *Brachyhyrax* from the Miocene, discovered at the latest Oligocene Losodok site (Rasmussen and Gutierrez, 2009). During the Neogene pliohyracids are known from the Miocene (23 Ma) to the Pleistocene (2.5 Ma). At that time, they flourished in Africa and Eurasia (i.e., in the Old World). The Old World pliohyracids were large, specialized herbivores with a heavy appearance. Many were the size of a tapir, but some became gigantic. Some were running species, and others were semi-aquatic. Pliohyracids have chalicothere-like teeth—that is, high-crowned (hypsodont), simplified, and sharp selenodont teeth adapted for eating leaves. Two species of the two Miocene genera *Afrohyrax* and *Meroehyrax* have been identified in the Oligocene Losodok locality (25–24 Ma) in Kenya. Fossils from the late Oligocene confirm the African origin of the pliohyracid family that colonized the entire Old World. The ancient endemic hyracoid lineages (Geniohyidae, Titanohyracidae, and Saghatheriidae) declined at the end of the Oligocene. In fact, the decline began earlier for the large hyracoid species due to competition with the anthracotheres that began to diversify in Africa during the early Oligocene. This is well seen in the Fayum series: In the upper levels of the Jebel Qatrani Formation, the anthracotheres increase in number, while the large hyracoids became rarer; conversely, in lower levels hyracoids are abundant, whereas anthracotheres are few (Rasmussen et al., 1992). After the Oligocene, all the Paleogene hyracoids, except for *Brachyhyrax*, became extinct and were replaced by pliohyracids and procaviids.

It was in the late Oligocene that the proboscideans were the most diversified in Island Africa (i.e., during the Paleogene), with the presence of seven genera. Stem groups such as barytheres and moeritheres had disappeared, leaving only the elephantiforms (i.e., proboscideans of modern elephantine type). The two early Oligocene palaeomastodont genera from the Fayum were still present and occurred in relative abundance in the late Oligocene sites. *Phiomia* is represented at Chilga (Ethiopia) by a huge species, *Phiomia major.*

There are also major new appearances of proboscideans in the late Oligocene. One is the genus *Chilgatherium*, found in Chilga (27 Ma) by William J. Sanders from the Museum of Michigan (Sanders et al., 2004). *Chilgatherium* was identified as the first-known deinothere. Deinotheres are the sister group of the elephantiforms. They are large elephantine-like proboscideans that flourished throughout the Old World from 23 Ma until 1 Ma, when they went extinct. *Chilgatherium* is indeed direct evidence to the origin of the deinotheres in Island Africa. The first Paleogene occurrence of the Miocene deinothere *Prodeinotherium* was reported from the latest Oligocene site of Losodok in Kenya (25–24 Ma). Another important appearance in the late Oligocene is *Eritreum*, which was discovered in Eritrea (Dogali, 27 Ma) by Jeheskel Soshani, founder of the Elephant Research Foundation (Shoshani et al., 2006). *Eritreum* is the earliest-known proboscidean with horizontal tooth replacement—that is, the earliest elephantoid, the group that includes the true mastodons (Mammutidae) and the extant family Elephantidae and all close relatives. *Eritreum* is a medium-sized proboscidean, with a shoulder height of about 1.30 meters.

Other remarkable appearances at the end of the Oligocene are the first precursors of the two major branches of modern elephantoids (or elephantimorphs): (1) true mastodons (mammutids) with lophodont teeth, represented by the genus *Losodokodon* from Kenya (25–24 Ma); and (2) gomphotheres with bunodont teeth (Elephantida), with *Gomphotherium* present as early as 27 Ma at Chilga. The gomphotheres are the stem group of the extant family Elephantidae. In the Miocene, they made the first evolutionary diversification of the Elephantida lineage, and they spread throughout the Old World as far as Japan!

The embrithopods have been found in three late Oligocene African sites, but they remain poorly known. The most important discovery is that of the giant species aptly named *Arsinoitherium giganteum*, from Chilga site (Ethiopia) (Sanders et al., 2004). Its estimated body mass is 2 to 2.4 metric tons, which corresponds to an animal the size of the white rhinoceros or slightly larger. This species is known from skull and jaw fragments, including a nasal horn, various bones, and massive, isolated teeth. Some of its teeth reach 13 centimeters in height. The embrithopods disappeared at the end of the Oligocene.

About 10 species of African rodents are known from several late Oligocene sites. Most come from the Namibian sites of Eocliff of possible but debated late Oligocene age (see above). All these rodents are African hystricognaths (i.e., phiomorphs). They are ancestral species of the two ex- tant African families of cane rats (thryonomyids) **(see fig. 5.10A)** and dassie-rats (petromurids). One species from Namibia, *Tufamys woodi*, is even considered by Sallam and Seiffert (2020) to be an early dassie-rat (family Petromuridae). The phylogenetic trees of the rodents based on fossil and molecular data (extant species) indeed indicate that the two extant African families of cane rats and dassie-rats diverged about 23 Ma, near the Oligocene–Miocene boundary. A few species and genera of these late Oligocene rodents are known in earlier beds of the Fayum, but most of them are new. There is a gap in the late Oligocene African fossil record for the anomaluroid group of flying squirrels that is known in several early Oligocene African sites. Similarly, the hamster cricetid family is not yet known in the late Oligocene of Africa, even though it has been found earlier, in the early Oligocene of Oman (Taqah).

The primates known in the late Oligocene of Africa are few compared to the early Oligocene. Six species are described, almost all of them an- thropoids. *Lokonepithecus* from Kenya is the last-known parapithecid. It is the last representative of an African diversification of ancestral an- thropoids (stem group of both the platyrrhines and catarrhines) that flourished at the beginning of the Oligocene (sites of the Fayum, Taqah, etc.). *Lokonepithecus*, known by one mandible and a few teeth, is a small species closely related to the Fayum genus *Apidium*.

All the other late Oligocene anthropoids are modern catarrhines, of the group that includes Old World monkeys and hominoids. A remarkable discovery made in Arabia by Zalmout and colleagues (2010) is that of *Saadanius hijazensis* (family Saadaniidae) in fossiliferous levels dated to 29–28 Ma. The fossils of *Saadanius* comprise a partial skull preserving the face, upper jaw, and palate **(plate 16)**. *Saadanius* is a stem catarrhine more advanced than the Fayum propliopithecids. *Saadanius hijazensis* was an arboreal species of medium size close to the siamang, with an estimated body mass of 15–20 kilograms.

Other major discoveries of primates in the late Oligocene of Africa, around 25 Ma, are those of the first cercopithecoids (Old Word monkeys) and hominoids (apes and humans). *Nsungwepithecus* from the Rukwa site (25 Ma), in Tanzania, is the oldest and most basal Old Word monkey

(Stevens et al., 2013). Extant Old World monkeys, like macaques, have characteristic bilophodont teeth that bear two transverse cutting crests. These teeth are specialized for a folivorous diet. *Nsungwepithecus* is the only cercopithecoid discovered in the late Oligocene, and it remains poorly known by a single fragment of mandible with a tooth. *Rukwapithecus* from the same Rukwa site is barely better known, with a broken mandible; it is the oldest primate of the hominoid group of the apes and humans, together with the subcontemporaneous genus *Kamoyapithecus* from Losodok (Kenya). *Kamoyapithecus* shares several hominoid characters, including strong sexual dimorphism evidenced by the discovery of a very large canine of a male individual. *Kamoyapithecus* is a medium-sized hominoid. Apart from catarrhines, a single tooth of a small lorisiform primate (the group of the extant lorises and galagos) has been reported from the Namibian site of Eocliff.

Another group known in the late Oligocene of Africa is the anthracotheres (artiodactyls). The most frequent late Oligocene African anthracotheres is the bothriodontine *Bothriogenys*, which is known earlier, in the early Oligocene levels of the Fayum. The Lokone locality (Kenya) yielded another bothriodontine, *Epirigenys lokonensis*. This species is important as it is considered by Lihoreau and colleagues (2015) to be the closest-known anthracothere stem group of the hippopotamids (see chapter 5). The nearby Kenyan site of Losodok yielded a third late Oligocene anthracothere of the genus *Brachyodus*, which was previously known only from the Miocene. This is the earliest-known occurrence of *Brachyodus*. It should be emphasized that the late Oligocene African anthracotheres are, overall, still poorly known.

From the Oligocene to the Miocene, a new world for Africa: The Old World

The Oligocene–Miocene transition, from 23 Ma, marks the end of the isolation of Africa with the Africa-Eurasia continental collision. It led to a major dispersal event in Africa (geodispersal event), and to the birth of the modern Old World faunal realm. Our knowledge of the Oligocene–

Miocene transition in Africa has been greatly improved by important new paleontological discoveries from the late Oligocene, between 25–23 Ma, but the very early Miocene, in the key 23–20 Ma time interval, is still poorly known due to a lack of fossils. Consequently, the initial stage of the modernization of African faunas and of the establishment of Old World faunas remain obscure. Much remains to be discovered about the arrival of Eurasian groups and their early evolution in Africa.

The fossil localities known in Africa from the very beginning of the Miocene are indeed rare. The oldest is widely believed to be Nakwai, in Kenya, dated at 22 Ma. Interestingly, Nakwai has yielded mostly ancient native African lineages, such as hyaenodonts (pterodontines), ptolemaiids, hyracoids, catarrhine primates, phiomorph rodents (thryonomyids, diamantomyids), and anthracotheres. Most of them are known from the Oligocene, even from the early Oligocene (34–31 Ma) of the Fayum. These include three ancient hyracoid families (titanohyracids, geniohyids, and saghatheriids) and rodents. This gives the Miocene fauna of Nakwai an archaic character. There are, however, new important appearances at the Nakwai site. One is *Alophia metios*, the earliest-known Neogene Old Word monkey (cercopithecoid) (Rasmussen et al., 2019). However, cercopithecoids are native to Island Africa, where there are known since the Oligocene with *Nsungwepithecus gunnelli* from Rukwa, Tanzania.

The major new African appearance in Nakwai is *Mioprionodon hodopeus* (Rasmussen and Gutierrez, 2009) **(fig. 4.31)**. It is the first true carnivore (Carnivora) appearing in Africa, apart from a possible stem species from the Zellah deposit (Libya). *Mioprionodon* is close to the extinct family Stenoplesictidae. This family, known from the Oligocene of Asia and Mongolia, includes hyper- or mesocarnivorous species that are stem groups of the viverrid family of the living civets. *Mioprionodon hodopeus* is probably related to the origin of both the African viverrids and the Malagasy euplerids (Rasmussen and Gutierrez, 2009). *Mioprionodon* is the forerunner of the great wave of Eurasian dispersal that is evidenced later by the fossils around 20 Ma. Further discoveries of ancient carnivores in Africa are to be expected as molecular phylogenies place the origin of extant Malagasy carnivores (euplerids) in Africa as early as the late Oligocene, around 27–25 Ma.

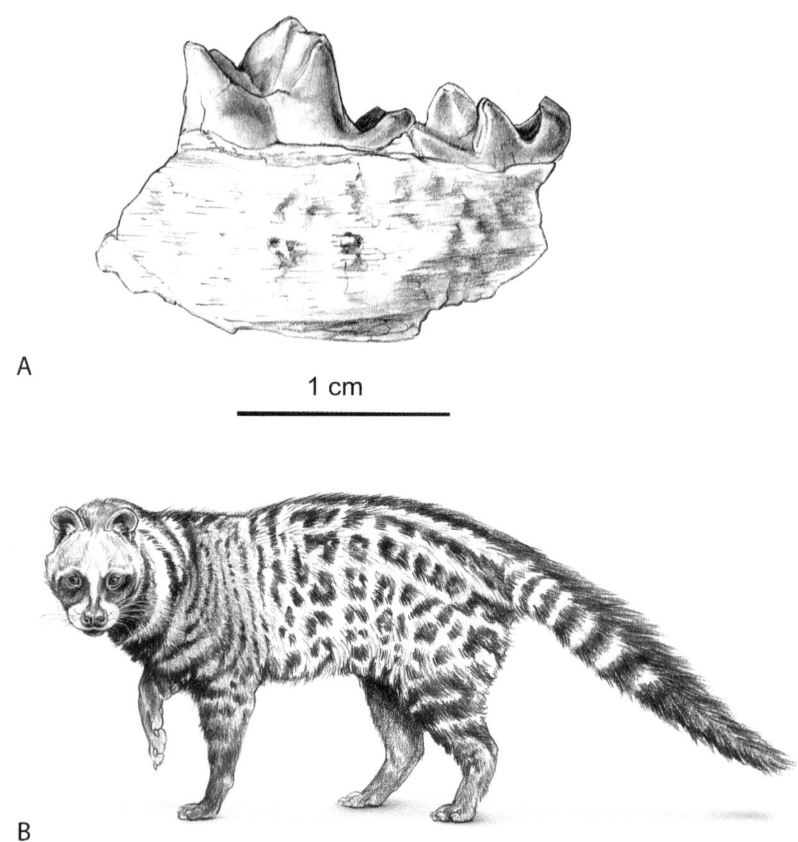

Figure 4.31. A. Partial lower jaw with the two last molars of the first true carnivoran appearing in Africa, *Mioprionodon hodopeus*, from the early Miocene of Nakwai, Kenya, 22 Ma. This species is known only from two lower-jaw fragments, the teeth of which have typical carnivoran morphology. *Mioprionodon hodopeus* is considered to belong to the stem group of both the African and Malagasy living civets and mongooses. **B.** The extant African civet *Civettictis civetta*, of the viverrid carnivore family, has evolved from early African stem groups such as *Mioprionodon*. Drawings by C. Letenneur.

This again implies gaps in the African fossil record of this key period of the Oligocene–Miocene transition, and it confirms that there is still much to be discovered about the early Eurasian colonization of Africa—we are still largely ignorant about the early episodes.

Other earliest African Miocene sites have been found in Kenya (Songhor, Koru, Meswa Bridge), Uganda (Napak, Moroto, Bukwa), and Namibia (Sperrgebiet). They have yielded old native African lineages (hyaenodonts, elephant shrews, aardvarks, lorisiform and hominoid primates, hystricognath rodents, proboscideans) but also many groups new to Africa. The aardvarks found in these sites (*Orycteropus, Myorycteropus*) are the earliest known, but their phylogenetic relationships indicate that they belong to an ancient African lineage (Tubulidentata, Afrotheria) that is not recorded during the Paleogene because of a large gap in the fossil record. These sites also yielded the earliest-known hippopotamids, in particular with *Morotochoerus ugandensis* (Orliac et al., 2010). According to the "anthracothere hypothesis" (see chapter 5), the hippos are not immigrant in Africa, but they evolved locally from an Oligocene African bothriodontine stem group (Lihoreau et al., 2015). All other new mammal groups found in these basal Miocene sites are evidence of the first wave of Eurasian dispersal into Africa, dated at 21–20 Ma. This corresponds to the first arrival in Africa of chalicotheres, rhinoceroses, giraffids, chevrotains, sanitheres, pigs (Suidae), genets and civets (Viverridae), mongooses (Herpestidae), extinct close relatives of felines (the Barbourofelidae), pinnipeds, pikas (sister group of the rabbits), various new rodents (including squirrels), and hedgehogs (erinaceids). After 20 Ma, several other waves of dispersals into Africa (immigrations) followed, with the successive arrival of different Eurasian groups. For example, various new rodents (hamsters, gerbils, moles, mice), felines, hyaenas, and bears arrived in a second wave, dating to 18–17 Ma based on several sites from Kenya (Rusinga, Kisingiri), Arabia, Egypt, Libya, and Namibia. Dogs, shrews, rabbits, porcupines, horses, and camels arrived in Africa only after the early Miocene (after 16 Ma).

The connection of Africa and Eurasia and the colonization of Africa by the Eurasian lineages led to an evolutionary competition with the endemic African fauna. There has been a coevolution of ancient and new lineages in Africa. This had two major consequences for the native African fauna: Some groups declined, whereas others recovered and experienced a remarkable evolutionary revival. The decline of the native African groups was progressive, not so strong at the beginning. The first disappearances at the beginning of the Miocene are those of stem anthropoid and stem catarrhine primates, such as the parapithecids, oligopithecids, propliopithecids, and of embrithopods and palaeomastodont proboscideans. Disappearances of the ancient families of hyraxes

(Geniohyidae, Titanohyracidae, and Saghatheriidae) was quickly followed by that of several Oligocene lineages of hyaenodont carnivorous mammals (Apterodontinae and Teratodontinae). The last native groups to disappear in Africa were the strange ptolemaiids by 12.5 Ma, the ultimate hyaenodonts at the end of the Miocene, and the pliohyracid hyraxes by 2.6 Ma (but these survived later in Europe and Asia).

The case of the African hyaenodonts is interesting. During the Miocene the hypercarnivorous lineage of the hyainailourines evolved into colossal predators, such as *Megistotherium*, *Hyainailouros*, and *Simbakubwa*. They were the largest-known terrestrial carnivores of the Cenozoic (together with the early artiodactyl *Andrewsarchus* from the Eocene of Asia). They specialized into a very particular carnivorous niche as a result of competition with true carnivores that arrived in the early Miocene and rapidly evolved successfully in Africa from that time on. These megapredators were, however, very dependent and ecologically vulnerable. Their disappearance at the end of the Miocene marks the final extinction of the hyaenodont mammals.

Besides these extinctions, several ancient native groups survived and became part of the living African fauna.

CHAPTER 5

Evolution and Relationships of the Major Endemic Groups in Island Africa

The afrotherians

Molecular phylogeny points to the existence of a major natural group (clade) of African placental mammals called the Afrotheria. It is one of the three first evolving branches of the placental mammals. It originated in Island Africa as early as the Cretaceous. It represents the main and oldest placental evolutionary radiation in Africa. Indeed, the clade Afrotheria comprises seven orders of African placental mammals—a third of the extant placental orders **(fig. 5.1)**. It includes small insectivore-like mammals (Afroinsectiphilia: tenrecs, golden moles, sengis, aardvarks) and large ungulate-like mammals (Paenungulata: dassies, sea cows, elephants, arsinoitheres). Afrotherian mammals have varied and quite different morphologies, sizes, and adaptations, which makes it difficult to reconstruct the ancestral Bauplan of the group as a whole. There is, for example, a debate whether the ancestral afrotherian was insectivore-like (e.g., plantigrade mammals with generalized locomotion, and having piercing-cutting teeth to feed on small animals) or ungulate-like (e.g., digitigrade mammals with cursorial locomotion, and having bunodont crushing teeth to feed on plants).

The early paleontological history of afrotherians is still poorly known—and even entirely unknown for some groups such as the aardvarks. Their fossil record is heterogeneous and incomplete, with long gaps before the Eocene. The first-known afrotherians in Africa are the stem paenungulates

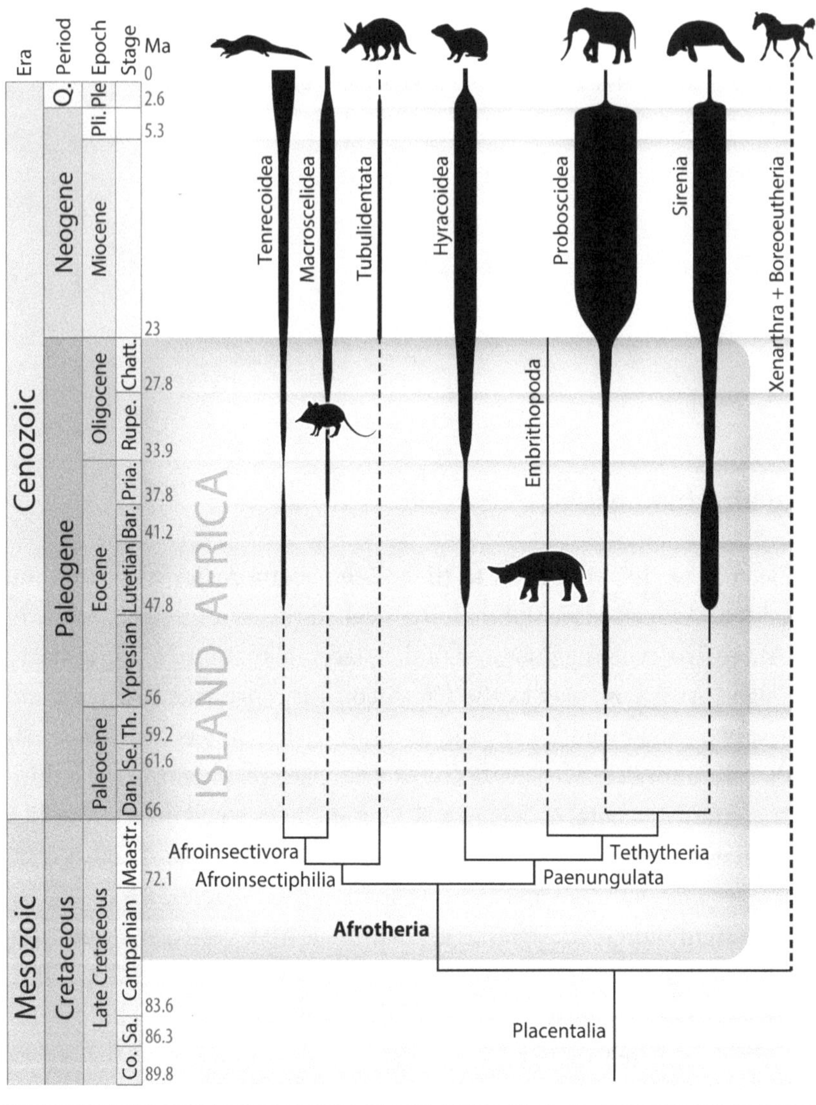

Figure 5.1. Relationships, stratigraphic distribution, and relative abundance of the afrotherian mammal orders. The thickness of the *bars* representing the afrotherian orders is proportional to their diversity (number of genera), which varies over time. It should be noted that from the Lutetian, the sirenians spread out from Africa and colonized the seas such as the Tethys, becoming cosmopolitan in their distribution. *Solid lines* represent known stratigraphic extension, while *dotted lines* represent possible temporal extension (not documented by fossils) inferred from the relationships. The line for the Xenarthra and Boreoeutheria represents the two other major clades that differentiated, along with Afrotheria, during the early evolution of placentals (their relationships are not shown here). Graph produced by A. Lethiers.

Hadrogeneios, *Abdounodus*, and *Ocepeia* and the earliest-known proboscidean, *Eritherium*, all from the middle Paleocene, circa 60 Ma, of Morocco (Ouled Abdoun Basin). There is also a possible tenrecoid afrotherian from the late Paleocene, circa 57 Ma, of Morocco (Ouarzazate Basin). These Paleocene mammals provide the best fossil evidence of the early African evolution of the Afrotheria. In particular, they indicate that the extant paenungulate orders of the elephants, dassies, and sea cows diverged at least at the beginning of the Cenozoic, and that the origin of the afrotherians goes back to the Cretaceous.

The known fossil record shows that the evolutionary radiation of afrotherians in Africa reached its peak between the end of the Eocene and the end of the Oligocene, with an explosion of shapes, sizes, adaptations, and species. This is especially true for paenungulates, which colonized most of the herbivore niches of Island Africa and evolved amazing and strange mammals such as the embrithopod *Arsinoitherium*. At the end of the Oligocene, by 23 Ma, when Africa's isolation was interrupted, major extinctions affected the afrotherians. Some entire groups, such as embrithopods, disappeared. Nevertheless, several afrotherian orders, such as the proboscideans, successfully survived the Oligocene–Miocene extinctions and underwent new and spectacular evolutionary radiations as they expanded throughout the Old World. The most marked decline of afrotherians occurred at the end of the Pliocene, by 2.6 Ma **(fig. 5.1)**. As a result, known living paenungulates appear to be relictual compared with their rich past abundance and diversity.

Insectivore-like afrotherians (Afroinsectiphilia): Golden moles, otter shrews, tenrecs, elephant shrews, and aardvarks

The origin and early evolution of the endemic African insectivores remain poorly known because of gaps for the early Cenozoic fossil record **(fig. 5.2)**. The fossil record of this group is much poorer than that of the paenungulates.

Apart from a possible Paleocene tooth, the earliest African relatives of the tenrecoid group (also called afrosoricidans)—which includes the golden moles (chrysochlorids), tenrecs (Malagasy tenrecines), and otter

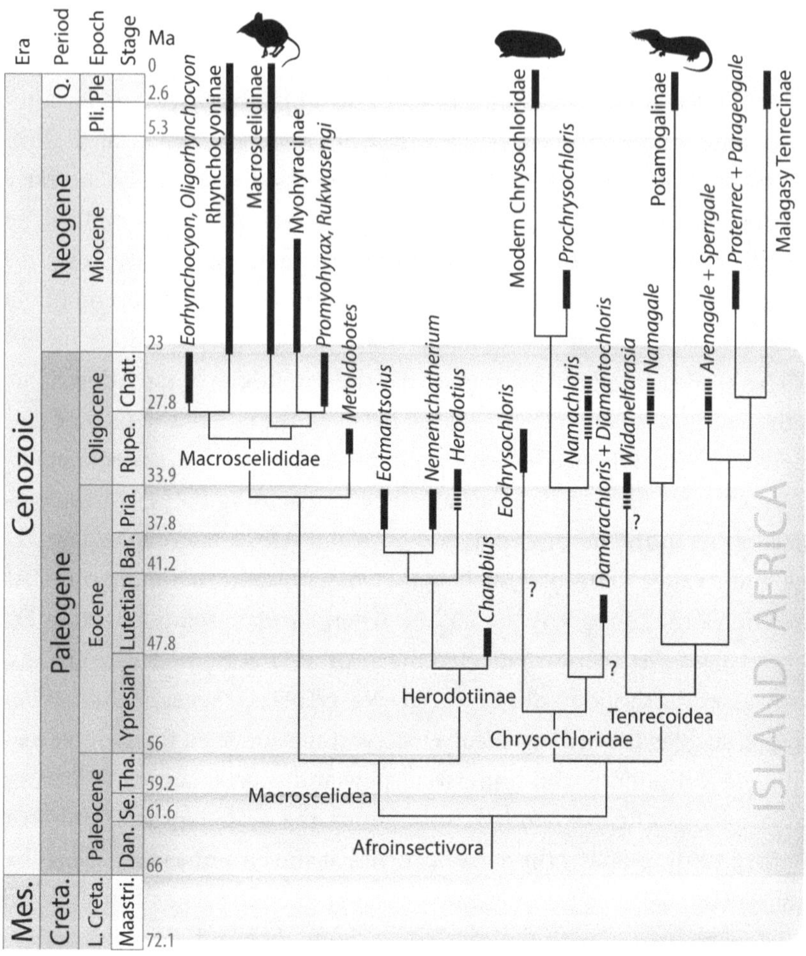

Figure 5.2. Relationships and stratigraphic distribution of afrotherian insectivore mammals (Afroinsectivora, Afrotheria). *Broken bars* represent uncertain stratigraphic extension for some clades (due to lack of data for dating the fossil sites). Graph produced by A. Lethiers.

shrews (potamogalines)—have been found in the middle Eocene of Black Crow, Namibia. Among them, *Diamantochloris* and *Damarachloris* have been initially described as the possible earliest relatives of the extant golden mole family (chrysochlorids). They share the tenrecoid zalambdodont puncturing teeth specialized for eating insects and worms, but in a poorly advanced state, indicating that if they really are chrysochlorids, they belong to basal and old lineages. More recent tenrecoids from the late Eocene (*Dilambdogale)* and the early Oligocene (*Eochrysochloris, Widanelfarasia*) of the Fayum have been reported, but they also remain much less specialized in dental morphology with respect to modern tenrecoid families. Conversely, *Namachloris arenatans* from the Eocliff site (Namibia) is an advanced tenrecoid closely related to the origin of the extant golden moles. *Namachloris arenatans* has been dated by Bronner and colleagues (2024) to circa 27 Ma, making it the earliest-known golden mole. It displays many characteristic specializations of the living golden moles in the teeth, skull, and postcranial skeleton (Bronner et al., 2024; Mason et al., 2018; Pickford, 2015d). In particular, *Namachloris arenatans* has a pneumatized skull base (i.e., bones with air cells), and its postcranial skeleton shows morphological traits "indicating a sand-swimming lifestyle as in extant golden moles" (Pickford, 2015d).

Other insectivore-like mammals from the Eocliff site have been identified as the earliest otter shrews (*Namagale*) and tenrecs (*Arenagale* and *Sperrgale*) (**fig. 5.2**), although their exact relationships within tenrecids remain to be established (Bronner et al., 2024). They also share the advanced zalambdodont dental morphology of the modern tenrecoids. Today, the otter shrews (potamogalines) are strictly African, and the tenrecs (tenrecines) are only known in Madagascar. The Oligocene Namibian tenrecines add to those found in the Miocene of Kenya, such as *Protenrec* and *Parageogale*, as paleontological evidence of the African origin of the Malagasy tenrecines, following a dispersal through the Mozambique Strait, in agreement with molecular studies of extant species. More generally, the fossils found in Namibia provide the best paleontological evidence for the early African evolution of the tenrecoid insectivores.

The African Paleogene fossil record of elephant shrews (macroscelidean group including the living *Elephantulus*) is patchy, which renders their phylogeny and evolutionary history poorly known and often debated. The earliest relatives of the elephant shrews (**fig. 5.3**) have been found in early–middle Eocene localities (ca. 48 Ma) of North Africa with the genus *Chambius*, which is the earliest and basal-most herodotiine. Other early herodotiine macroscelideans (*Nementchatherium, Metoldobotes, Herodotius*) are known in the late Eocene and early Oligocene of North Africa. They are very primitive macroscelideans with low, crushing (i.e., bunodont)

Figure 5.3. The extant African elephant shrew *Rhynchocyon* (Macroscelididae, Afrotheria) is a "giant" elephant shrew (body length 28 cm; body mass 500–700 g). Elephant shrews, also called sengis, are small insectivorous mammals strictly endemic to Africa; they are very fast saltatorial land mammals with a distinctive long and mobile snout (hence the name elephant shrew). Drawing by C. Letenneur.

teeth adapted to a herbivorous diet, resembling those of early paenungulates and some European "condylarths" (louisinids) more than modern (extant) elephant shrews, which are in contrast insectivorous. These herodotiines are early stem groups to the extant macroscelidid family, and the Fayum genus *Metoldobotes* (Metoldobotinae) is their closest and later relative. Herodotiines and metoldobotines disappeared in the late Oligocene, giving way to the evolutionary radiation of the modern elephant shrews (sengis), today represented by the family Macroscelididae, comprising the Miocene subfamily Myohyracinae and the two extant subfamilies Rhynchocyoninae and Macroscelidinae. The earliest relatives of extant rhynchocyonine subfamily were recently found in the late Oligocene of Tanzania (*Oligorhynchocyon*) and Namibia (*Eorhynchocyon*).

Modern elephant shrews are distinctive, endemic small mammals living in Africa. They are characterized by an elongated and mobile snout—hence the "elephant" in their name. They are among the fastest small mammals. They use their elongated and slender hindlimbs for running by rapid bounds. Extant sengis are both highly specialized with many original traits (e.g., they use their tongue to catch insects and

Figure 5.4. The living burrowing *Orycteropus afer* (body length 100–130 cm). This strictly African mammal is nocturnal and specializes in feeding on ants and termites. This is the only known living species of the afrotherian order Tubulidentata. Aardvarks are thought to have originated in the Late Cretaceous, by 75 Ma, but their oldest fossils date to only 21 Ma. Drawing by C. Letenneur.

worms, they are saltatorial and monogamous, and they have small litters of precocial young) and adapted to a wide variety of environments in Africa, from the coastal deserts to montane forests (Rathbun, 2009). They are especially well diversified in the rocky and forested habitats of Africa. Their adaptation to varied African environments allowed them to survive competition with the Laurasian insectivores and ungulates that dispersed to Africa in the Miocene. They even underwent a moderate diversification illustrated by almost 20 extant species in Africa.

The ancient evolution in Island Africa of the aardvarks (Afrikaans word meaning "earth pigs") **(fig. 5.4)**, of the tubulidentate order, is entirely unknown. No fossils are known in Africa before 21 Ma (early Miocene). This large gap in the fossil record is possibly related to the peculiar mode of life of aardvarks, which are solitary, nocturnal, fossorial, anteaters (myrmecophagous). Living aardvarks have been related to paenungulates, such as proboscideans and hyracoids, based on similar morphology of vertebra, phalanges, hands, feet, and muscles. In particular, they share the unguligrade stance. In addition, the bilobed outline of their teeth suggests an ancestral pattern with four tubercles (quadritubercular plan) as in the paenungulates. Current molecular phylogeny, however, relates the aardvarks to insectivore-like afrotherians (Afroinsectiphilia). This suggests that aardvarks diverged from other insectivore-like afrotherians around 75 Ma, in the Late Cretaceous. Some recent paleontological studies have suggested that the enigmatic ptolemaiid mammals from the Oligocene of the Fayum might be an extinct sister group of the tubulidentates (Gunnell et al., 2010; Seiffert, 2007). Ptolemaiids are small, extinct insectivore placentals endemic to Africa. Their specialized teeth resemble those of aardvarks, with high and bilobed crowns, but may be convergent. The recently extinct and enigmatic "Malagasy aardvark," *Plesiorycteropus*, was included in a separate order (Bibymalagasia), distinct from the tubulidentates. Recent studies confirm that *Plesiorycteropus* is an afrotherian, but its exact relationship to the African tubulidentates remains a subject of debate. In any case, the afrotherian relationship of the aardvarks indicates they have an ancient African origin: There is in Africa a long ghost lineage for tubulidentates, likely extending into the Late Cretaceous. With only one living species, the aardvark order (Tubulidentata) is today the least diverse of living placental mammals, although during the Miocene, the tubulidentates underwent a modest evolutionary radiation in the Old World that is documented by about 10 species.

Ungulate-like afrotherians (Paenungulata): Hyraxes, embrithopods, sea cows, and elephants

Paleogene paenungulates are well documented in Africa, especially the proboscideans and hyraxes. The earliest and basal-most hyracoid is *Seggeurius*, from the Ypresian of Algeria (51 Ma) and Morocco (56 Ma). This is the only known early Eocene hyracoid. Another slightly more recent stem hyrax is *Microhyrax*, from the middle Eocene of Algeria (48 Ma). These basal hyraxes are small and have primitive morphology with bunodont and poorly specialized lophodont and selenodont molars, and simple premolars. In addition to *Microhyrax*, a remarkable diversity of other hyracoids, which are more specialized, has been uncovered in the middle Eocene sites (48 Ma) of Algeria (Gour Lazib) and Tunisia (Chambi). They include at least seven species of the two families Geniohyidae and Titanohyracidae. They represent the first-known hyracoid evolutionary radiation with species of varied morphology and specializations in their locomotion and diet. The adaptive diversity of these early hyraxes is surprisingly close to that of the Fayum communities. It points to the antiquity of the evolutionary history of the order Hyracoidea, more so than the few known earlier fossils, such as *Seggeurius*. It is likely that *Seggeurius*—like *Microhyrax*—was the relict of an ancient basal lineage of hyrax **(fig. 5.5)**.

The initial diversification of hyracoids known in the middle Eocene Algerian and Tunisian sites originated at least at the beginning of the Eocene, and possibly during the Paleocene. As a result, we must expect new significant fossil discoveries in African beds of these ages. From the late Eocene to the end of the Oligocene, hyracoids became the most diversified paenungulates. There are about 30 hyracoid species belonging to the three known endemic African Paleogene families (Titanohyracidae, Geniohyidae, and the paraphyletic "Saghatheriidae"). They illustrate the peak of the endemic adaptative radiation of the hyracoids in Africa. They evolved many convergences in locomotion and diet with Laurasian euungulates. Bovid-like hyracoids (*Antilohyrax*), piglike hyracoids (geniohyids), and brontothere-like hyracoids (titanohyracids) have evolved. Some species, such as *Titanohyrax ultimus*, were gigantic, with a body mass of about one metric ton. *Titanohyrax ultimus* probably evolved independently from earlier giant hyracoids, such as "*T.*" *mongereaui* from the middle Eocene. It should be noted that the early and middle Eocene history of the hyracoid radiation is very poorly known because of important gaps in the fossil record in Africa during this period. Following the arrival of euungulates in Africa, the three endemic Paleogene hyracoid families declined rapidly after the Oligocene. They have been replaced by two new Neogene lineages: the extinct pliohyracid family and the living procaviid family.

Procaviids are descendants of the Oligocene stem genus *Thyrohyrax* (see, for example, Barrow et al., 2012). The relationships of the Miocene large-bodied pliohyracid hyraxes remain poorly known. In some hypotheses pliohyracids are also related to the Paleogene genus *Thyrohyrax*, and in others they are the sister group of the titanohyracids (Barrow et al., 2012; Seiffert, 2007) **(fig. 5.5)**. Pliohyracids possessed high-crowned (hypsodont) and selenodont teeth, and some were specialized to a semiaquatic

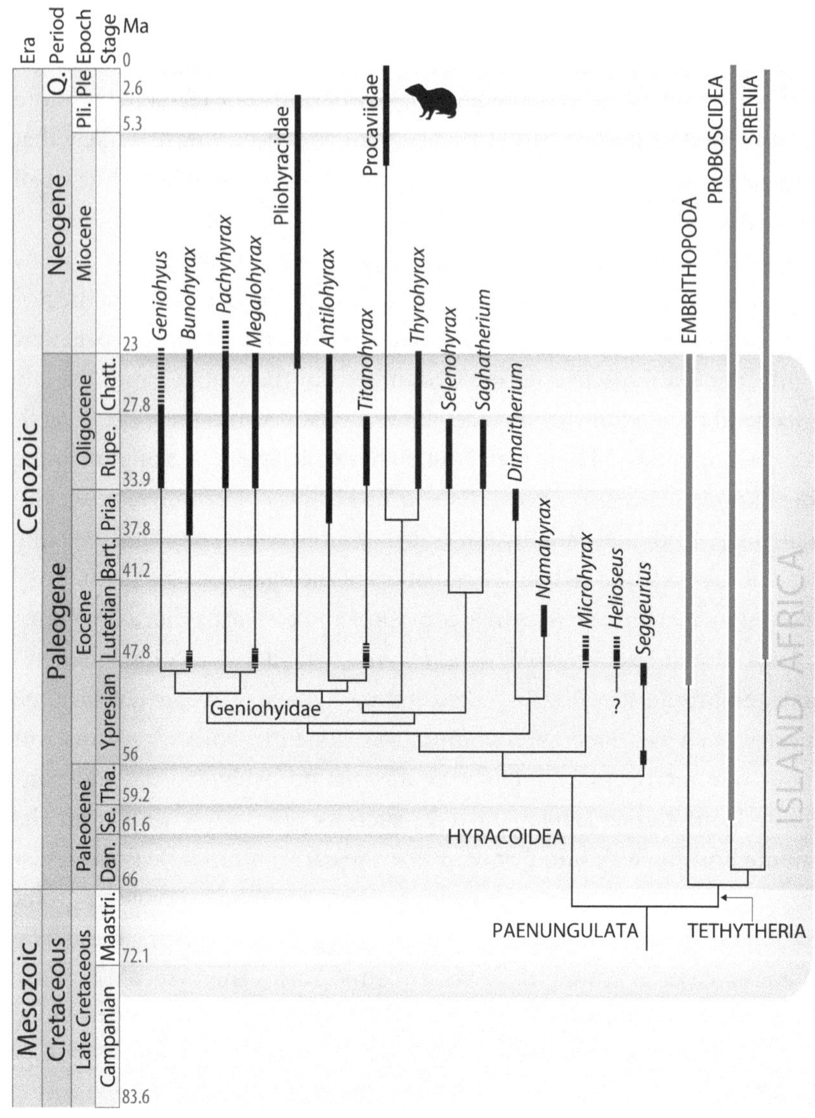

Figure 5.5. Relationships and stratigraphic distribution of the hyracoids (Hyracoidea, Afrotheria). *Broken bars* represent uncertain stratigraphic extension for some clades (due to lack of data for dating the fossil sites). Graph produced by A. Lethiers, modified from Barrow et al. (2010).

life. They appeared in Africa at the end of the Oligocene, and they made a modest diversification during the Miocene–Pliocene with seven genera that expanded in the Old Word. This is the third radiation known in the evolutionary history of the hyracoids. Pliohyracids became giant, with a body mass up to 1,400 kilograms, during the Pliocene–Pleistocene. They went extinct in Africa at the beginning of the Pleistocene, by 2.6 Ma.

The last and most discrete radiation of the hyracoids is that of the extant dassies family, the Procaviidae, which appeared in the late Miocene by 10 Ma. Living procaviids are known by three genera and five species. With respect to euungulates, the procaviids occupy a peculiar ecological niche with distinctive adaptations (e.g., they are good jumpers and climbers), habitats, and feeding habits, which explains their survival in living African faunas.

Phylogenetic studies, especially those based on morphological and paleontological data, suggest that elephants, sea cows, and embrithopods are more closely related to each other than to hyracoids. They belong to a group called Tethytheria, a name that means "beasts of the Tethys Sea" (McKenna, 1975). For a long time, it was assumed that the forebears of the tethytheres were semiaquatic mammals that lived and evolved on the African shores of the Tethys Sea. An alternative phylogenetic hypothesis to the clade Tethytheria, supported by some molecular studies, is that hyraxes and elephants are more closely related to each other than to sea cows.

The extinct order of embrithopods first appeared in the early Eocene of Morocco with the genus *Stylolophus*. *Stylolophus* is much more primitive and smaller than the spectacular and highly derived Oligocene genus *Arsinoitherium*. However, it shares the strange and unique hyperdilambdodont teeth (bearing very wide W-like crests) of *Arsinoitherium*. Interestingly, the smallest and most ancient species, *Stylolophus minor*, is larger, by twice the body mass, than the coeval basal proboscidean *Phosphatherium*. Early embrithopod species increased in size fast. The larger and slightly younger species *S. major* (53 Ma) illustrates the early trend, at the beginning of the Eocene, toward an increase of the paenungulate's body size, also seen in proboscideans. *Stylolophus* indicates that embrithopods were an early offshoot of the modern tethytheres, having evolved before the origin (i.e., divergence) of the proboscideans and sirenians. In addition, *Stylolophus* confirms the African origin of the embrithopods and their ancient evolution extending back to the beginning of the Cenozoic. *Stylolophus* is close in morphology to the palaeoamasiids known in the Eocene and Oligocene of Anatolia and Romania. It indicates that palaeoamasiids evolved from an African *Stylolophus*-like ancestor that crossed the Tethys sometime between the early Eocene and the early–middle Eocene transition, by 56 to 48 Ma.

The medium to large-sized embrithopod *Namatherium* from the middle Eocene (ca. 43 Ma) of Namibia is more advanced and closer to *Arsinoitherium* in skull and tooth morphology, although it is smaller and less specialized. It shares a deep palate for housing a large and prehensile tongue and teeth that become more lophodont-like with an ectoloph (external crests) that is transformed into two long transverse crests, in the same way as in *Arsinoitherium*. *Namatherium* is a close ancestral relative of *Arsinoitherium*, both belonging to the family Arsinoitheriidae (Gheerbrant et al., 2025). Although the roof of the only known skull of *Namatherium* is broken, there are some morphological indications that it may have had nasal horns as in *Arsinoitherium*. *Arsinoitherium* from the Oligocene is the best-known and most spectacular paenungulate known from Island Africa. It was an impressive mammal with a gigantic size, close to that of the rhino, and huge nasal horns. It evolved a graviportal skeleton, as in proboscideans and other large mammals, with limbs arranged upright into vertical columns beneath the body, and lophodont teeth convergent with other paenungulates. The embrithopods remained discrete throughout their evolution with a modest diversity. Only two or three species of the genus *Arsinoitherium* are known from the Oligocene, the time of their apogee. However, these species were abundant in the Fayum localities, as evidenced by their numerous fossils. *Arsinoitherium* increased in size in the late Oligocene to elephantine size with the well named species *A. giganteum*. The embrithopods went extinct at the end of the Oligocene, by 23 Ma.

The earliest-known proboscidean, *Eritherium*, was found in the middle Paleocene (60 Ma) phosphate beds of Morocco. It is an archaic and very small proboscidean, the size of extant dassies. It lacks all spectacular features of the extant elephants, such as the trunk and tusks. However, it shares morphological details that are only known in proboscideans, such as the construction of the orbit (with maxillary participation) and the dental morphology (molar bunodont and incipient lophodont, first lower incisor enlarged). *Eritherium* is close to the origin of the proboscideans (**fig. 5.6**). It represents a very early evolutionary stage rarely fossilized in modern placental orders, and particularly in highly specialized orders such as that of the modern elephants. *Phosphatherium*, from the early Eocene (56–55 Ma), is more derived than *Eritherium* and a little larger, resembling in size a medium-sized dog. This is the first proboscidean with specialized lophodont teeth. It had a folivorous (leaf-eating) diet, which probably evolved in relation to the climatic warming and vegetation changes that occurred at the beginning of the Eocene.

During the early and middle Eocene there was a rapid increase in body size and a multiplication of lophodont proboscidean lineages with folivorous diets. The first large proboscideans, which were tapir sized, are the genera *Daouitherium* (53 Ma) and *Numidotherium* (51 Ma). The Eocene–Oligocene proboscideans became major components of the earliest mammal megafauna of Africa, together with the embrithopods and some hyrax species. The first diversification of the proboscideans during the Eocene was not that of a single group (clade), but it was the result of a rapid sequential (paraphyletic) succession of several basal lineages showing a mosaic evolution of advanced traits. It is represented by the successive lophodont lineages of the phosphatheres, daouitheres, numidotheres, and barytheres, which evolved in an ancient favorable folivorous herbivore niche in Island Africa.

Barytherium (**see fig. 4.16**), known from 37 to 27 Ma, is the first proboscidean of elephantine size. Together with *Arcanotherium*, it is the last representative of the evolutionary sequence of the lophodont proboscideans that precedes the evolution of modern elephantiforms. Although *Barytherium* is a large proboscidean, its postcranial skeleton remained primitive, with limbs still like that of *Numidotherium* (Delmer, 2005). It does not have columnar limbs (graviportal posture) but flexed limbs and plantigrade hands and feet of a cursorial animal. However, *Barytherium* is more advanced than *Numidotherium* in several traits,

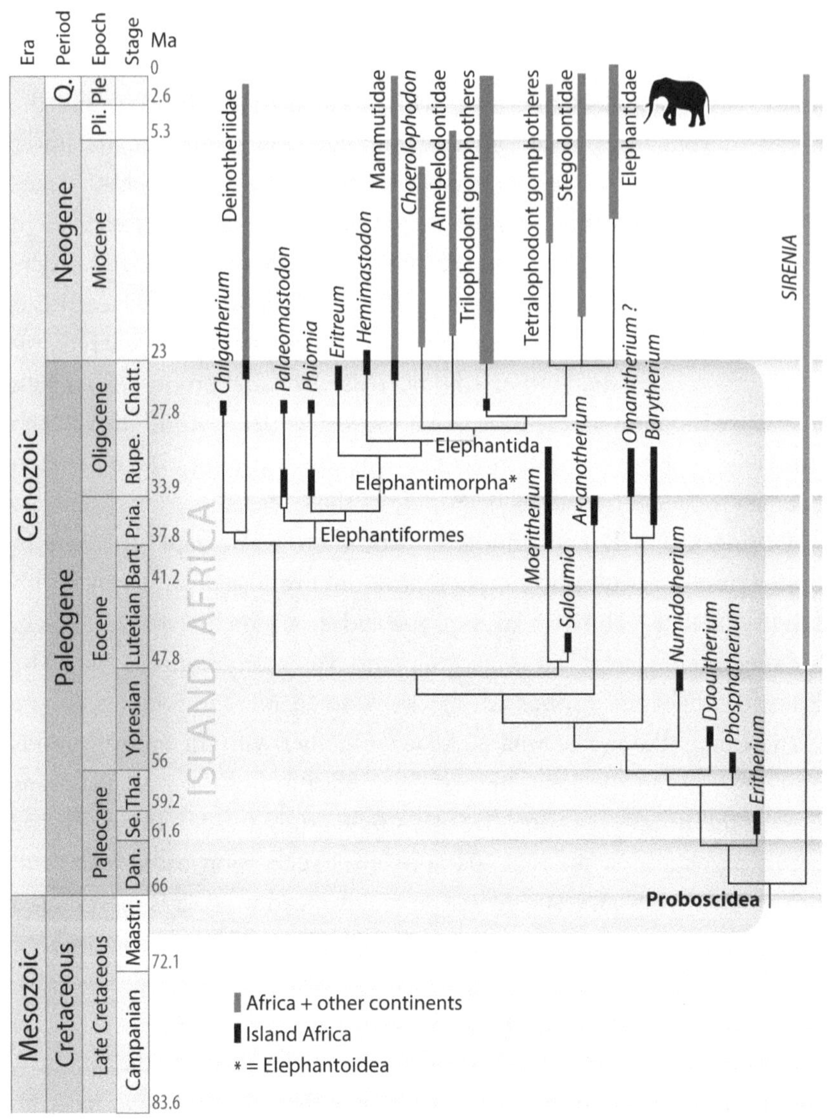

Figure 5.6. Relationships and stratigraphic distribution of early relatives of elephants (Proboscidea, Afrotheria). Graph produced by A. Lethiers, modified from Gheerbrant and Tassy (2009).

such as a more anterior orbit, the loss of some teeth, and larger incisors. *Arcanotherium* is particularly interesting in the evolution of modern proboscideans (i.e., elephantiforms) as it is intermediate in morphology and relationship between the basal barytheriid proboscideans with cutting lophodont teeth and the modern proboscideans with crushing bunolophodont teeth, such as *Moeritherium* and elephantiforms.

It was during the middle Eocene that the first proboscideans with bunolophodont crushing teeth appeared: the moeritheres. Moeritheres evolved as a peculiar offshoot of basal proboscideans that eventually went extinct. In a remarkable parallelism, during the late Eocene and early Oligocene they followed the adaptive evolution of the sirenians with the semiaquatic genus *Moeritherium*, which fed on aquatic plants. *Moeritherium* has an elongated body carried by short legs, as in the stem sea cows and hippos. *Moeritherium* evolved specialized bunodont-lophodont teeth in which the tubercles swell and multiply at the expense of the primitive cutting crests, giving them a distinct crushing function. This announced the diet of modern elephantiform proboscideans of elephantine type that have crushing bunolophodont teeth. The moeritheres are indeed an early lateral lineage detached from the root of the elephant-like elephantiforms. Like *Arcanotherium*, early moeritheres from the Eocene show a dental morphology typically intermediate between the Eocene lophodont proboscideans with cutting teeth and *Moeritherium* and Oligocene elephantiforms with crushing teeth.

The later proboscideans of elephantine appearance are the deinotheres and the elephantiforms, which are sister groups. They both have the graviportal stance, tusks, and a trunk. Deinotheres are more generalized than elephantiforms in the elongated skull with a smaller trunk. They are also distinguished by lophodont cutting teeth, an absence of upper tusks, and lower tusks curved downward. Their lophodont teeth are probably secondary with respect to stem proboscideans. The earliest-known deinothere, *Chilgatherium*, occurred in the late Oligocene of Africa. It is basal to the Old Word deinotheres, which evolved from the Miocene to the Pliocene. During this 20-million-year-long period, deinotheres evolved with a modest diversity and without significant changes, except in size. They were evolutionarily conservative throughout their existence. This means they were well adapted to a particular and stable ecological niche.

The first modern elephantiform proboscideans appeared in the Oligocene (34 Ma) with the palaeomastodonts *Palaeomastodon* and *Phiomia*. They evolved key advanced features of modern elephants, such as ever-growing tusks, bunolophodont teeth with three lophs, an elephantine trunk, and a pneumatized skull (in which the bones are hollowed out to lighten the weight). The palaeomastodonts disappeared at the end of the Oligocene, by 23 Ma. They were replaced by the elephantoid group (also called elephantimorphs), which includes true mastodons (Mammutida, the lophodont elephantoids) and elephantidans (Elephantida, the bunodont elephantoids, such as gomphotheres and extant elephants). The first-known and basal-most elephantoid is *Eritreum* from the late Oligocene. *Eritreum* shares the most remarkable specialized traits of modern elephants, such as the horizontal tooth succession and the enlarged skull, tusks, and trunk. More advanced elephantoids of the gomphothere group, which diversified and spread widely across the world during the Miocene, appeared in Africa in the late Oligocene. The most remarkable evolutionary radiation in the history of the proboscideans is that of the elephantoid. It is at the origin of the extant elephantid family. The elephantids made a limited radiation during the Miocene–Pliocene. The three living elephant species are the relics of a rich and flourishing ancient evolution. This is why we can say that the diversity of proboscideans is almost entirely past.

Paleontology also documents an important evolutionary radiation of the sea cows (sirenians) during the Eocene. However, among paenungulates the sirenian radiation differs from the others by the fact that it occurred outside continental Africa, in the Tethys Sea, which the sirenians colonized during the Eocene from the African coast. The early diversification of the sea cows followed their adaptation to a major new ecological niche, the niche of marine herbivorous mammals. Earliest and basal-most sirenians belong to the prorastomid family, which is well known in western Tethys.

Prorastomids are semiaquatic stem sirenians, but they are more specialized semiaquatic mammals than, for instance, the hippos. The evolutionary transition of the sirenians from their ancestral terrestrial condition to their derived semiaquatic adaptations remains entirely unknown due to the lack of Paleocene and early Eocene fossils. The paenungulate relationship of the sirenians and recent discoveries in the early Eocene of Tunisia and Senegal indicate they have an African ancestry. The sirenian semiaquatic adaptation probably evolved in ancestral species that lived in close ecological relation with fresh waters of the lakes and rivers of Island Africa during the Paleocene and early Eocene times. Indeed, it is known that during the Eocene huge lakes occupied great inland basins in northern and eastern Africa, in particular at the place of the former interior trans-Saharan sea **(see fig. 2.3)**.

The later evolutionary radiations of semiaquatic and fully aquatic sirenians are documented in the marine Paleogene localities of North Africa. They include several stem groups of sirenians, such as prorastomids, protosirenids, and the paraphyletic "halitheriines." The stem sirenians went extinct at the end of the Oligocene. They were replaced by the Miocene and Pliocene diversification of the extant families of the dugongs (dugongids) and manatees (trichechids). The sea cows colonized the Tethyan Sea as far as the West Indies in the middle Eocene. One middle Eocene species of *Eotheroides* has also been reported in marine deposits of Madagascar (Samonds et al., 2009).

The African rodents

Two major ancient endemic groups of rodents are known in Island Africa: the anomaluroids and the hystricognaths (also called hystricomorphs). They are at the origin of several extant African lineages, some very distinctive. The anomaluroids gave rise to the living flying scaly-tailed squirrel (*Anomalurus*), flying mouse (*Idiurus*), and flightless scaly-tailed squirrel (*Zenkerella*). The hystricognaths of the African phiomorph group gave rise to living cane rats, dassie-rats, and mole-rats. These two major rodent groups experienced a successful and broad adaptive radiation in Africa. Some lineages dispersed out of Africa during the Paleogene. A third rodent group, the hamster family (cricetids, muroid rodents), appeared much later in Africa, during the Oligocene, following a dispersal from Asia.

Anomaluromorphans: Origin of the flying scaly-tailed squirrels and springhares

Anomaluroids are the earliest-known rodents in Africa. They appeared at the beginning of the middle Eocene, by 48 Ma, with several genera of the paraphyletic (unnatural) family "Zegdoumyidae," which is basal to all anomaluroids **(fig. 5.7)**. The zegdoumyids are the basal-most stem groups of living families of (1) the anomalurids, which include the flying scaly-tailed squirrel (*Anomalurus*) and flying mouse (*Idiurus*); and (2) the zenkerellids, which include the flightless scaly-tailed squirrel (*Zenkerella*). The zegdoumyids include several genera, such as *Glibia*, *Glibemys*, and *Zegdoumys*, and they evolved up to the early Oligocene with *Dakhlamys*. The zegdoumyids are in fact very specialized, especially in their teeth, indicating that their African evolution is earlier than the middle Eocene and might be at least as old as the early Eocene.

A younger and more evolved group of stem anomaluroids, the paraphyletic "Nementchamyidae," appeared in the late Eocene. Its origin in Africa was also probably earlier (i.e., middle Eocene). By the late middle Eocene–late Eocene the nementchamyids dispersed out of Africa into South Asia, where they gave rise to *Pondaungimys* from the latest middle Eocene of Myanmar. The nementchamyid *Kabirmys* was the largest anomaluroid known in Island Africa; it had the size of extant species of *Anomalurus*, with a body length of 20–30 centimeters. The earliest representatives of extant anomaluroid families appeared in late Eocene and early Oligocene African sites. *Oromys* and *Prozenkerella*, from the early Oligocene (34 Ma), are the earliest-known zenkerellids. The oldest species of the extant African genus *Zenkerella* has been found in the earliest Miocene. The oldest anomalurids appear in the late Eocene (37 Ma) with *Shazurus* from the Fayum-BQ2 site (Egypt), followed by *Argouburus* and *Paranomalurus* from the early Oligocene of Dakhla (Morocco) (Marivaux et al., "Anomaluroid Rodents," 2017). Extant anomalures (anomalurids and zenkerellids) are nocturnal and arboreal rodents living in African

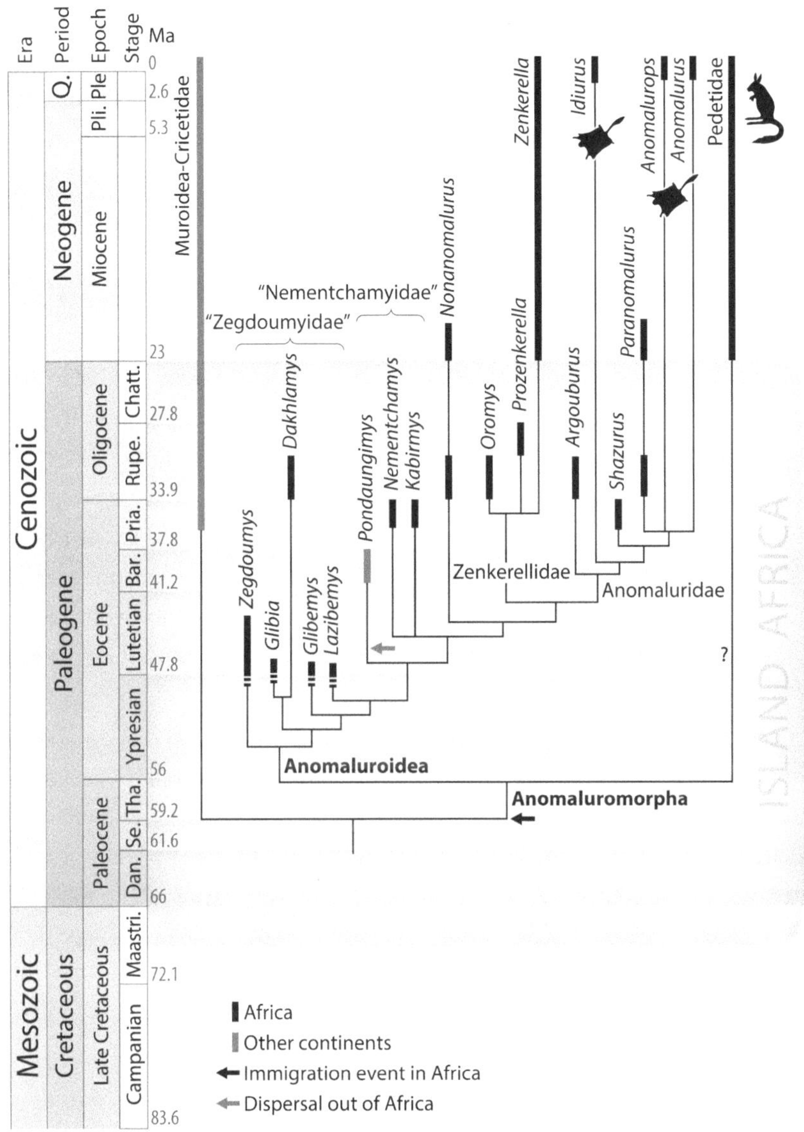

Figure 5.7. Relationships and stratigraphic distribution of the anomaluroid rodents (Anomaluroidea, Rodentia). "Zegdoumyidae" and "Nementchamyidae" are in quotation marks because they are paraphyletic groups. *Broken bars* represent uncertain stratigraphic extension for some clades (due to lack of data for dating the fossil sites). Graph produced by A. Lethiers, modified from Marivaux et al. ("Anomaluroid Rodents," 2017).

forests. All are gliding rodents **(see fig. 4.27)**, except for *Zenkerella*, which is the only living nonvolant anomaluroid. Molecular phylogeny has shown that the nongliding genus *Zenkerella* is sister group to all anomalurids, which means that the gliding adaptation evolved only once in modern anomalures (Fabre et al., 2018).

Molecular phylogeny indicates that the springhares, of the extant African family Pedetidae, are the nearest living relatives of the extant anomaluroids, both forming a clade called Anomaluromorpha (Montgelard et al., 2008). They share, for instance, the hystricomorph morphology of the upper jaw. However, pedetids are unknown before the Miocene (23 Ma), and the origin of this strange endemic family remains entirely enigmatic. Pedetids are highly specialized nocturnal African rodents. They resemble both rabbits and kangaroos. They are terrestrial and saltatory bipeds **(fig. 5.8)**, adapted to relatively open environments, in contrast to the arboreal anomalurids that live in forested environments. In addition, recent studies discovered that the nocturnal springhares are biofluorescent mammals (i.e., they have fur that glows under ultraviolet light) (Olson et al., 2021).

Given the divergent and highly specialized morphology and adaptations of the anomalures (arboreal, gliding) and springhares (terrestrial, saltatorial), and the fact that extant anomalure families (anomalurids and zenkerellids) are known since the late Eocene, these two endemic rodent lineages must have diverged very early in Africa, at least by the middle Eocene (48 Ma) and probably earlier. This agrees with molecular phylogeny that dates the phylogenetic separation of the anomalurids and zenkerellids to at least the beginning of the Eocene, by 56 Ma (Montgelard et al., 2008). Consequently, pedetids and anomaluroids diverged very early in Africa, and there is a long ghost lineage of more than 30 Ma for the pedetids. Another great challenge for paleontological field research in Africa!

Hystricognaths: Origin of the cane rats, dassie-rats, and mole-rats

The second major endemic group of rodents present in Island Africa is the hystricognaths **(fig. 5.9)**. Hystricognaths appear in Africa nearly

Figure 5.8. The living nocturnal African springhare rodent, *Pedetes capensis* (Anomaluromorpha, Rodentia). This strange rodent looks like a cross between a kangaroo and a rabbit. It is a large (body length 35–45 cm) and specialized saltatorial terrestrial rodent endemic to Africa. It has an ancient African common ancestor with the African flying squirrels (anomalures), dating to at least the early Eocene (about 55 Ma), but for which we still have no known fossil. The earliest-known springhares come from the early Miocene of Kenya, Uganda, and Namibia. Drawing by C. Letenneur.

10 million years later than the anomaluroids, at the end of the middle Eocene by 39 Ma, with the species *Protophiomys tunisiensis*. This is coincidental with the appearance of anthropoid primates in Africa (Marivaux et al., "New and Primitive Species," 2014). However, *Protophiomys tunisiensis* is nearly contemporaneous with the earliest-known South American caviomorphs, and the estimated time of origin of the hystricognaths was probably somewhat earlier, about 45 Ma (middle Eocene).

Hystricognaths are represented in Africa by phiomorphs, such as the living cane rats, dassie-rats, and mole-rats (**fig. 5.10**) and all their early relatives. In South America hystricognaths are represented by the caviomorphs, such as the living agoutis, chinchillas, and guinea pigs and early relatives. All these rodents have a hystricognath mandible in which the angle between the horizontal and vertical branches of the mandibles is strongly shifted laterally with respect to the vertical plane of the incisor. Their upper jaw is hystricomorphous: It is characterized by an enormous infraorbital foramen through which passes a large masticatory muscle (the medial

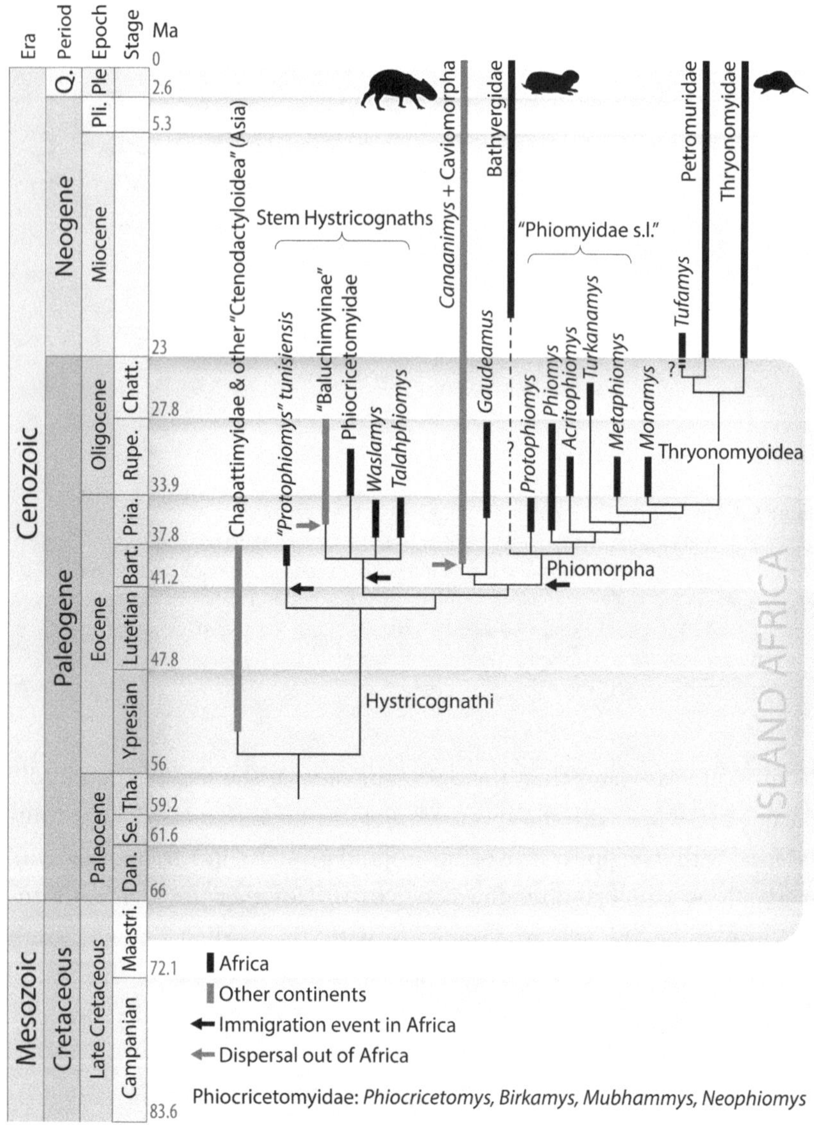

Figure 5.9. Relationships and stratigraphic distribution of the hystricognath rodents (Hystricognathi, Rodentia): cane rats, dassie-rats, and mole-rats. "Phiomyidae" is in quotation marks because this is a paraphyletic group. A *broken bar* represents uncertain stratigraphic extension for some clades (due to lack of data for dating the fossil sites). Graph produced by A. Lethiers, modified from Marivaux et al. ("Earliest Oligocene," 2017).

Figure 5.10. Two living African hystricognath rodents. **A.** The cane rat, *Thryonomys* (family Thryonomyidae), is a large diurnal rodent (body mass 4–7 kg; body length 35–61 cm) that lives in marshes and close to rivers and lakes, where it feeds on aquatic grasses. **B.** The naked mole-rat, *Heterocephalus glaber* (family Bathyergidae), also called blesmols, is a distinctive and highly specialized burrowing rodent of diminutive size (body length 8–10 cm; body mass 50–80 g) living in sub-Saharan Africa. The cane rat and naked mole-rat share an unknown Eocene African ancestor. Drawings by C. Letenneur.

masseter, one of the main chewing muscles of mammals) to attach to the snout. This arrangement of the muscle favors the horizontal chewing movement. The enamel of the teeth of the hystricognaths has a distinctive microstructure called multiserial. Hystricognaths have taken much greater advantage of the Island Africa paleoecosystems and paleoenvironments than the anomaluroids. They exploded in diversity and became one of the most diverse and abundant mammals in the African Paleogene fossil localities.

The African hystricognaths are related to Asian stem groups such as ctenodactyloids. They arrived in Africa from Asia during the same episode of dispersal that included the anthropoids. This dispersal across the Tethys Sea involved an ancestor close in age and morphology to *"Protophiomys" tunisiensis* (Marivaux et al., "New and Primitive Species," 2014). According to paleontologists like K. Christopher Beard (University of Kansas), the colonization of Africa from Asia by both the hystricognath rodents and the anthropoid primates was done by oceanic drifting on plant rafts across the Tethys Sea. It took place during the climatic warming period of the Middle Eocene Climatic Optimum (the MECO; Scotese et al., 2021), around 41 Ma. This global event was marked by intense monsoons and floods that resulted in large rafts of vegetation being washed down rivers and deltas in equatorial Asia and carried into the Tethys Sea. These vegetation rafts formed floating islands that allowed the dispersion of small mammals, such as rodents and primates, during the Eocene (Beard et al., 2017) and, subsequently, their evolutionary success in new areas such as Africa.

After their arrival in Island Africa, the hystricognaths rapidly diversified by the end of the Eocene and the Oligocene. A series of successive stem groups, which are ancestral to extant families of mole-rats, cane rats, and dassie-rats, evolved. The basal-most African hystricognaths found in the late Eocene are *Protophiomys*, *Talahphiomys*, *Waslamys*, and *Phiocricetomys*. One of these stem lineages returned to Asia during the late Eocene, giving rise to the baluchimyines. During the Oligocene, the "phiomyid" heterogeneous family included stem groups closer to the extant phiomorph families. The early Oligocene genus *Gaudeamus* (34 Ma) tells another story. It has peculiar specializations shared with the South American caviomorph rodents. *Gaudeamus* is indeed a basal caviomorph, and it provides direct fossil evidence of the African origin of the South American caviomorphs. The earliest caviomorph known in South America has been found earlier, in a late middle Eocene site of Peru, dated at 41 Ma (Antoine et al., 2012). The divergence of African phiomorphs and South American caviomorphs must therefore be earlier than *Gaudeamus,* dating back from the middle Eocene. This is consistent with molecular studies that have estimated the caviomorph-phiomorph separation at about 45–41 Ma.

The divergence of the cane rats (thryonomyids) and dassie-rats (petromurids) is estimated to be close to the Oligocene–Miocene boundary, around 23 Ma. Phiomorph rodents also include the highly derived family Bathyergidae, which comprises the living African mole-rats, also called blesmols. Molecular phylogeny indicates that mole-rats are the sister group to both the extant cane rats and dassie-rats (i.e., to the thryonomyoids) and that they diverged close to the Eocene–Oligocene boundary, at about 36 Ma. Mole-rats are distinctive burrowing rodents. They include about 20 living species, all found in sub-Saharan Africa. The most remarkable of them is the naked mole-rat, *Heterocephalus*, which is included in its own family, Heterocephalidae, by Patterson and Upham (2014); this African rodent is cold-blooded (ectothermic), and it has exceptional longevity. The paleontological origin of these extant African families of phiomorphs remains poorly known, especially for the bathyergids. *Monamys* from the early Oligocene is the closest stem relatives of the thryonomyoids. The earliest fossils of extant families thryonomyids, petromurids, and bathyergids are early Miocene (23–20 Ma), and the earliest-known bathyergid is *Renefossor* from Napak and Songhor, about 20 Ma. The exact stem group of the bathyergids is unknown; it is probably a generalized Oligocene "phiomyid" close to *Phiomys*, *Birkamys*, and *Mubhammys* (Sallam and Seiffert, 2020). Several genera from the late Oligocene or early Miocene of Namibia, such as *Tufamys*, are tentatively identified as early dassie-rats (petromurids).

The last major group of rodents known in Island Africa is the muroid hamster family (Cricetidae). It is known only in the early Oligocene of Oman by a few teeth. This is its first appearance in Africa. It arrived in Africa following a recent dispersal from Eurasia.

The extinct African carnivorous mammals: Hyaenodonts

Hyaenodonta (previously included within the polyphyletic "Creodonta") is the only group of carnivorous mammals known in Island Africa (apart from a few uncertain carnivoran identifications). This is an extinct group of placental predators that was successful not only in Africa but also in Laurasia during the Paleogene. It probably colonized Laurasia from Africa during the Paleocene **(fig. 5.11)**, and it disappeared from Laurasia at the end of the Oligocene because of competition from true carnivores (Carnivora). In Africa, it diversified successfully during the Paleogene in the absence of true carnivores, and it survived later than in Laurasia, until the middle Miocene. Most hyaenodonts were generalized terrestrial predators, but some lineages were remarkably specialized.

Hyaenodonta first occurs in Africa in the middle Paleocene (60 Ma) with the koholiine *Lahimia*. Hyaenodonta appears a little later in Asia, during the late Paleocene, with the Limnocyoninae. Koholiines are an early endemic African lineage that evolved during the Paleocene and early Eocene. They include *Lahimia* (Selandian of Morocco, 60 Ma), *Boualitomus* (Ypresian of Morocco, 56 Ma), and *Koholia* (Ypresian of Algeria, 51 Ma). They remain poorly known, and their relationship within the hyaenodonts is uncertain. Koholiines have intriguing precociously specialized features, such as the loss of the first premolar. This supports an early African origin of the hyaenodonts at the beginning of the Paleocene. This is consistent with the explosive adaptive radiation of placentals at this time and, in particular, with the rapid colonization of the carnivorous niche in Africa made vacant after the KPg extinctions. There is additional evidence supporting the African origin of the Hyaenodonta. The most important evidence is the fossils of *Tinerhodon* from the late Paleocene of Morocco. Although more recent than the koholiine *Lahimia*, *Tinerhodon* is more generalized and basal, representing an ancient relict lineage. It is either the basal-most known hyaenodont or a stem relative of the European proviverrines. In the late Paleocene Adrar Mgorn fossil site (Morocco, 57 Ma), the co-occurrence of *Tinerhodon* with insectivore-like basal placentals such as *Cimolestes cuspulus* (family Cimolestidae), which are possible generalized stem hyaenodonts, is also in agreement with hyaenodonts' African origin. In addition, there is evidence of an intriguing diversity of unidentified hyaenodontid-like isolated teeth from this site (Gheerbrant, 1995).

The koholiines went extinct after the early Eocene. They gave way to a new and large endemic group of hyaenodonts, which underwent a unique

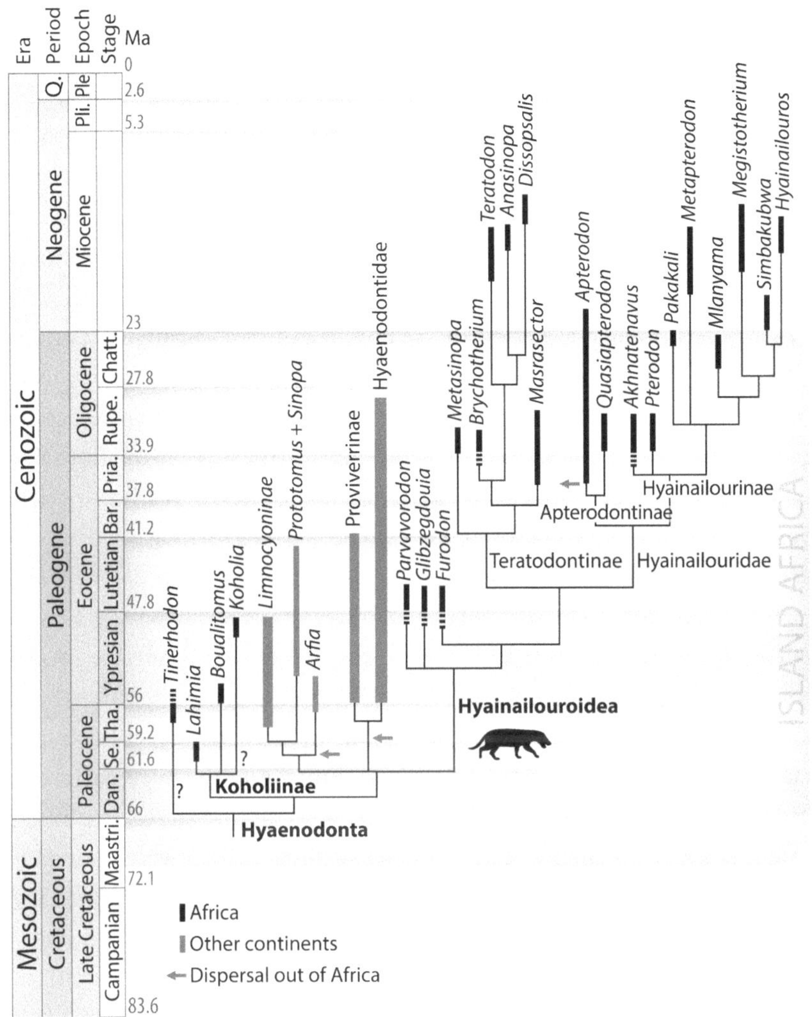

Figure 5.11. Relationships and stratigraphic distribution of the extinct African hyaenodont carnivorous mammals (Hyaenodonta). *Broken bars* represent uncertain stratigraphic extension for some clades (due to lack of data for dating the fossil sites). Graph produced by A. Lethiers, modified from Solé et al. (2021).

diversification without competition in Island Africa: the Hyainailouroidea. By the middle Eocene, the hyainailouroids underwent a first evolutionary radiation in Island Africa with the sudden appearance of six species (Solé et al., 2014) of the two subfamilies Hyainailourinae and Teratodontinae. An additional hyainailouroid lineage appeared a little later, in the late Eocene, and became rapidly successful: the apterodontines. The middle–late Eocene hyainailouroids evolved varied morphologies unknown in earlier times in Africa. This resulted from a successful African evolutionary radiation in the absence of true carnivores (extant order Carnivora). The hyainailouroid diversity also increased in concert with that of the small African herbivorous mammals, such as hyracoids and rodents, on which they fed. Hyaenodonts with varied adaptations occurred in the Eocene with arboreal, semiaquatic, and terrestrial species. Most of them are small, although the hyainailourines include large hypercarnivorous species the size of a fox or a wolf.

The hyaenodonts reached their evolutionary peak during the Oligocene in Island Africa. Earlier hyainailouroid lineages continued to evolve and diversify. All three Eocene lineages—hyainailourines, apterodontines, and teratodontines—were successful in the Oligocene, but especially the semiaquatic *Apterodon* (apterodontine) and the hypercarnivore *Pterodon* (hyainailourine). *Pterodon* was among the largest mammal predators living in Island Africa. Apterodontines, which were abundant in the Oligocene, are distinctive hyainailouroids adapted to a semiaquatic lifestyle. Their very robust tooth shape indicates that they fed on hard-shelled prey, such as crustaceans, crabs, and shrimps. Apterodontines colonized Europe in the Rupelian as indicated by the lineage formed by the two related species *Apterodon macrognathus* (Africa) and *Apterodon gaudryi* (Europe). Well-preserved fossils of the teratodontine *Masrasector* (late Eocene and early Oligocene) recently found in the Fayum (Borths and Seiffert, 2017) indicate it was a small, skunk-sized generalist terrestrial carnivorous mammal with a varied diet, including small vertebrates, insects, fruits, and nuts. It was particularly fond of the small terrestrial mammals abundant in the same localities, such as the hyracoids and the phiomorph rodents. With only four known species, the diversity of hyainailouroids is much lower during the late Oligocene, but this is probably related to gaps in the fossil record at that time. The three hyainailouroid lineages of the hyainailourines, apterodontines, and teratodontines were indeed present in the late Oligocene. The hyaenodonts are among the major African endemic mammal lineages that survived the closure of the Tethys in the Oligocene–Miocene transition.

Miocene hyaenodonts include the apterodontines, teratodontines, and hyainailourines. Apterodontines rapidly went extinct in the early Miocene, followed by teratodontines in the late Miocene. The hypercarnivorous hyainailourine lineage successfully evolved during the Miocene and specialized as giant predators such as *Megistotherium*, *Hyainailouros*, and *Simbakubwa*. *Megistotherium* reached the size of a rhinoceros. Some of these hyainailourine megapredators, such as *Hyainailouros*, colonized Eurasia in the Miocene. All hyaenodonts disappeared in the late Miocene.

Primates: Origin of the loris, galagos, lemurs, monkeys, and apes

The earliest-known primate in Africa is *Altiatlasius* from the late Paleocene (about 57 Ma). It is the first-known "true" primate (Euprimates) in the world, together with *Altanius* from the late Paleocene of China. *Altiatlasius*'s relationship within the primates (**fig. 5.12**) remains uncertain and very debated because it is only known by a few isolated teeth. Some paleontologists argued it is the oldest-known anthropoid and that it is the fossil evidence of the possible African origin of anthropoids. However, the hypothesis cannot be excluded that it was a short-lived basal euprimate lineage that soon went extinct in Island Africa. *Altiatlasius* itself is likely of Asian origin. It indicates an ancient primate colonization of Island Africa, during the Paleocene if not earlier.

The strepsirrhine primates found in Island Africa are represented by a few extinct Laurasian immigrant adapiforms and by several endemic stem and crown strepsirrhines related to extant lorises and lemurs.

The extinct adapiforms (Adapidae) first colonized Africa from Eurasia in the middle Eocene with the caenopithecine *Namadapis*. A few other

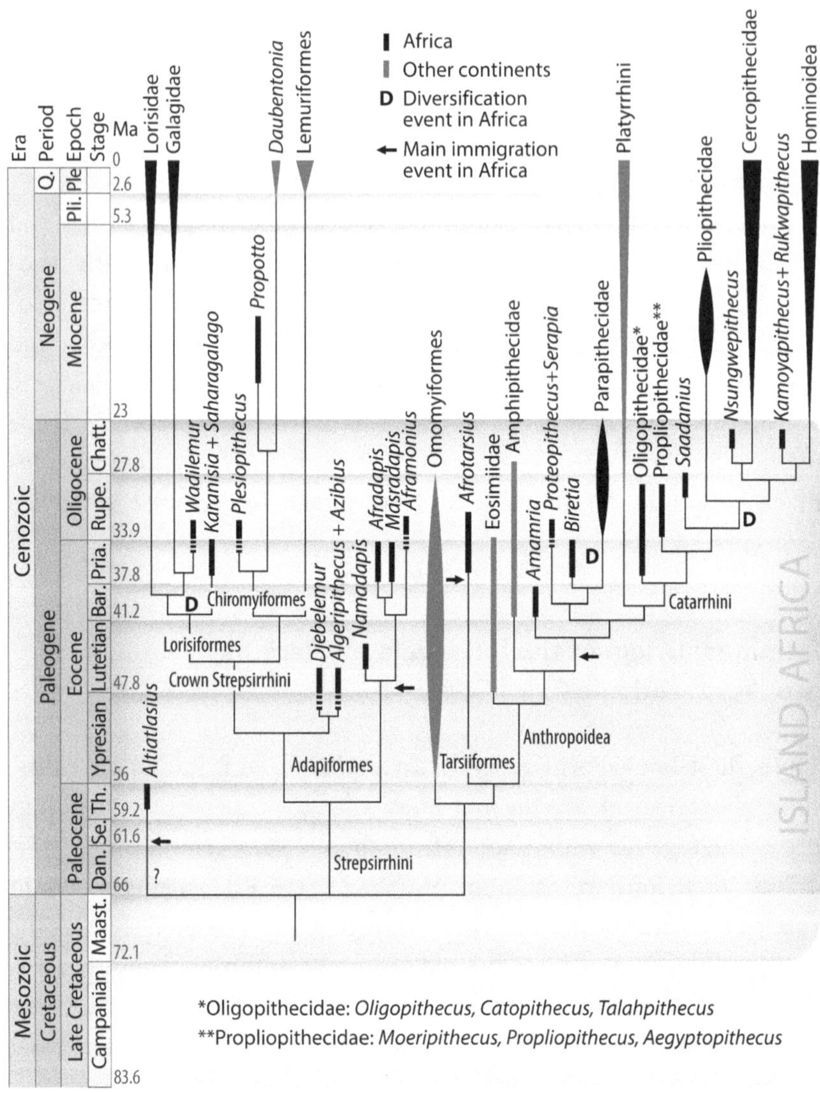

Figure 5.12. Relationships and stratigraphic distribution of the primates, with details of early primates known in Island Africa. Note that the strongly reduced diversity of extant Hominoidea is not illustrated here. *Broken bars* represent uncertain stratigraphic extension for some clades (due to lack of data for dating the fossil sites). The thickness of the *bars* varies with the taxonomic diversity of the groups, where thicker one indicate periods of particular evolutionary success for certain groups. Graph produced by A. Lethiers.

caenopithecines, such as *Afradapis and Masradapis*, are known during the late Eocene and the Oligocene. They are short-lived lineages in Africa. Their presence in Africa is either the result of a small endemic diversification since the middle Eocene or of other dispersal events from Asia or Europe across the Tethys Sea. The African adapiforms are remarkable in their convergent evolution with anthropoids. This is especially true of the folivorous genus *Aframonius*, but also of *Afradapis*. The adapiforms soon went extinct in Island Africa, before the end of the Oligocene. They likely disappeared because of competition from anthropoids, which evolved rapidly and successfully in Africa by the late Eocene and the Oligocene.

The earliest relatives of the modern strepsirrhines are found in Island Africa. They occur in the middle Eocene sites of Algeria and Tunisia with *Algeripithecus* (azibiid family) and *Djebelemur* (djebelemurid family). They are stem groups that evolved before the divergence of lemurs (lemuriforms) and lorises (lorisiforms). They lack the specialized tooth comb of modern strepsirrhines, which is a tool for feeding on tree gum and for grooming the coat. The first-known primates with a tooth comb are *Karanisia* and *Saharagalago*, from the late Eocene (Priabonian, 37 Ma), and *Wadilemur*, from the early Oligocene (34 Ma), all found in the Fayum. *Wadilemur* is the earliest galago (bush babies of the galagid family). *Saharagalago* (37 Ma) is either an earlier galago than *Wadilemur* or a stem lorisiform. *Karanisia* (37 Ma) is probably a stem lorisiform **(fig. 5.12)**. *Wadilemur* indicates that the two extant lorisiform families of the Galagidae and Lorisidae diverged in Africa before the Oligocene (i.e., before 34 Ma). *Plesiopithecus* from the early Oligocene of Egypt (Fayum-L41 site, ca. 34 Ma) shares unique specialized traits of the strange extant aye-ayes, such as a rodent-like mandible bearing a very large and anteriorly inclined incisor. It is indeed the earliest-known relative of the Malagasy aye-aye group, the Chiromyiformes. The Miocene *Propotto* was recently re-identified as another aye-aye from Africa (Gunnell et al., 2018). Both *Plesiopithecus* and *Propotto* are fossil evidence of the African origin of the Malagasy aye-ayes. Molecular studies have suggested that the common ancestor of Malagasy lemurs originated from Africa in the Paleocene or early Eocene, and that lemuriforms and chiromyiforms diverged in Madagascar after the arrival of their African common ancestor. However, the exact dating and the number of dispersals from Africa to Madagascar remain the subject of debate. In fact, the discovery of *Plesiopithecus* supports, instead, independent dispersals of stem lemurs (lemuriforms) and stem aye-ayes (chiromyiforms) to Madagascar (Gunnell et al., 2018) in the Oligocene (by 28 Ma) and early Miocene (by 20 Ma), respectively. The fossils thus tend to suggest that Madagascar was colonized twice by African strepsirrhines. It is worth noting that the Oligocene to early Miocene interval (34–20 Ma) is a dispersal period of several other mammal and vertebrate lineages to Madagascar. No fossil of true lemurs (lemuriforms) is identified in the African island, although some poorly known Paleogene primates, such as *Omanodon*, have interesting resemblances.

The presence in Africa of relatives of the tarsiiforms, the group of the Asian living tarsiers, is based on the afrotarsiid *Afrotarsius* known from the early Oligocene of the Fayum and the late Eocene of Libya. However, the relationships of *Afrotarsius* are hotly disputed. The tarsiiforms' *Afrotarsius* relationship is supported by dental and postcranial fossils (Rasmussen et al., 1998; Simons and Bown, 1985), and by comparison with Miocene fossils of tarsiids from Thailand (*Tarsius sirindhornae*) (Seiffert, 2012; Seiffert et al., 2018). Another school favors a possible relationship of *Afrotarsius* to the stem eosimiiform group of the anthropoids (Beard, 1998, 2016; Jaeger et al., "Late Middle Eocene," 2010), which are known from the middle Eocene of Asia. In both hypotheses, *Afrotarsius*, as a haplorrhine primate, is important for the discussion of the origin of African anthropoids (Godinot et al., 2018).

Amamria from the middle Eocene (Bartonian, 41 Ma) depicts the first appearance in Africa of the anthropoid primates (Marivaux et al., "Morphological Intermediate," 2014). It indicates that anthropoids colonized Africa in the middle Eocene from an Asian cradle. The arrival of the stem Asian anthropoids in the African island was likely the result of a dispersal on floating islands (i.e., by rafting). These floating islands were generated more frequently by the monsoon rains during the MECO period, between 41 and 40.5 Ma (Beard et al., 2017).

During the late Eocene (37 Ma), the anthropoids remained discrete in Island Africa. They are represented by two species of *Biretia*, which is a small basal African anthropoid. *Biretia* is more evolved than *Amamria*. It is the oldest parapithecid (parapithecoid). For instance, it shares with the Fayum parapithecines the bunodont teeth with large hypocone to crush the fruits on which they fed. It is, however, clearly more basal than the Oligocene parapithecines with distinct traits such as unfused mandibles.

After the Eocene, the anthropoids underwent a rapid and spectacular diversification in Island Africa with the evolution of about 20 species in four families. Four main Oligocene lineages of anthropoids evolved. The proteopithecids (*Proteopithecus* and *Serapia*) are a short-lived sister group to the parapithecids. Together with the proteopithecids, the parapithecids represent an evolutionary radiation of extinct stem anthropoids The oligopithecids and propliopithecids are stem catarrhines (basal to Old World monkeys and apes). All these anthropoids share many skeletal traits, such as bunodont crushing molars, orbit closed at the back, and fused frontals. They also show sexual dimorphism in traits such as canine size.

The proteopithecids are more generalized than the parapithecids, especially in tooth morphology, and they probably originated before the late Eocene. The parapithecids are small, generalized arboreal anthropoids. During the Oligocene they underwent a remarkable radiation that includes six genera (five from Africa and another one in South America) in the two subfamilies parapithecines and qatraniines. Parapithecids are ancestral to all living anthropoids—that is, to both the platyrrhines (New World monkeys) and catarrhines (Old World monkeys and apes). They are indeed more primitive than the catarrhines in several features, such as the presence three premolars in each half jaw instead of two. Parapithecids have bunodont crushing teeth (i.e., crown robust with swollen tubercles and crests) and were frugivorous. The thick enamel of their teeth indicates that they also crunched nuts and seeds. A parapithecid, *Ucayalipithecus*, was recently discovered in a latest Eocene or early Oligocene site in Peru (Seiffert et al., 2020). *Ucayalipithecus* is closely related to the early Oligocene African parapithecid *Qatrania*. *Ucayalipithecus* colonized South America following a dispersal across the Atlantic in the late Eocene, probably on natural rafts produced by tropical storms and carried by ocean currents and winds. The last-known parapithecid is *Lokonepithecus* from the latest Oligocene. Parapithecids went extinct at the dawn of the Miocene, probably in relation to the rapid diversification of the modern cercopithecoids at that time.

According to a now classical hypothesis, the prehensile-tailed New World monkeys (platyrrhines) originated from an Eocene African anthropoid stem group that remains unknown. The molecular clock indicates that the New World monkeys (platyrrhines) and the Old World monkeys and apes (catarrhines) diverged in the middle Eocene, between 44 Ma and 40 Ma. This is slightly earlier than the earliest fossil record of both the platyrrhines (*Perupithecus)* and catarrhines (*Talahpithecus*?; *Catopithecus*), which is late Eocene (37 Ma) or early Oligocene (34 Ma). The platyrrhine colonization of South America therefore followed a transatlantic dispersal from Africa, at some point between the middle and late Eocene (i.e., between 44 Ma and 37 Ma). Current studies indicate that the colonization of South America by the platyrrhines was the result of a transatlantic dispersal independent from that of the parapithecids (Seiffert et al., 2020).

The first catarrhines appeared in Africa during the late Eocene or the early Oligocene. They are represented by the two families oligopithecids and propliopithecids, which are basal branches that evolved before the rise of the Old World monkeys (cercopithecoids) and the apes (hominoids: lesser apes, great apes, and humans). They have only two premolars per half jaw, like all the extant catarrhines, but they remain very primitive—for example, in the presence of a small brain and unfused mandibles. The oligopithecids and propliopithecids are small arboreal quadrupedal anthropoids, and their orbital size indicates they were diurnal. The oligopithecids are the basal-most catarrhines: They are less derived and probably more ancient than the propliopithecids. Oligopithecids are small early catarrhines primates with a body mass up to one kilogram. They have cutting teeth with large mortar-like basins to crush and grind the food. These teeth indicate they had a more varied diet than the propliopithecids, consisting of fruit, leaves, and insects.

Propliopithecids are more advanced, apelike early catarrhines than oligopithecids. Three genera are known in the Oligocene: *Propliopithecus,*

Moeripithecus, and *Aegyptopithecus*. They are medium-sized generalized catarrhines with body masses ranging from one to eight kilograms. Among them *Aegyptopithecus*, which has a body mass of five to eight kilograms, is the largest anthropoid primate known in the Fayum faunas. The propliopithecids have crushing bunodont teeth with four large tubercles very similar to those of generalized catarrhines from the early Miocene of East Africa. These teeth indicate anthropoids with a typically frugivorous diet. *Aegyptopithecus* shows strong sexual dimorphism, especially in the canine, the skull, and the body. The skull displays a strong sagittal crest and a more elongated face in males than in females (**see fig. 4.30**). It is evidence that propliopithecids lived in small polygamous social groups that were organized around a dominant male. Propliopithecids disappeared in the late Oligocene.

Saadanius (family Saadaniidae) from 29–28 Ma old beds (early–late Oligocene) of Arabia is more evolved than the Fayum propliopithecids and much larger, with a body mass of 15–20 kg. It is the most advanced known stem group of the Old World monkeys and apes (Catarrhini). The first modern ("crown") catarrhines appeared at the end of the Oligocene, about 25 Ma. This is *Rukwapithecus* and *Nsungwepithecus* from the Rukwa site (Tanzania)—which are the earliest ape (hominoid) and the earliest Old World monkey (cercopithecoid), respectively—and also *Kamoyapithecus* from Losodok (Kenya), which is another early hominoid contemporaneous with *Rukwapithecus*. Although still poorly known, these three modern catarrhines testify to a divergence of the Old World monkeys (cercopithecoids) and apes (hominoids) in Africa during the Oligocene (Stevens et al., 2013) consistent with molecular dating. Old World monkeys (cercopithecoids) and apes (hominoids) rapidly diversified in the Miocene (with first the Miocene record by 22 Ma), and they spread from Africa throughout the Old World.

The anthropoid primates are one of the major ancient mammal lineages of Island Africa that survived the closure of the Tethys and successfully evolved until recent times, expanding well beyond Africa.

Artiodactyls: Relationships of the African anthracotheres and hippopotamid origin

Relationships of the anthracotheres are controversial, with two competing hypotheses. The classical and long-held hypothesis, based on comparative anatomy, is that the African anthracotheres belong to an extinct lineage that originated from a Eurasian bothriodontine stock, and that hippopotamids stem independently from a basal artiodactyl group that includes peccaries and pigs (i.e., the suoids) (Gentry and Hooker, 1988; Kron and Manning, 1998; Pickford, 2008). However, molecular data indicate that hippos are not suoids (their resemblances being convergences) but are instead sister group of whales (Cetacea), forming the Whippomorpha (or Cetancodonta) clade. Consistent with this hypothesis, a morphological and paleontological study by Jean-Renaud Boisserie's team from the University of Poitiers (Boisserie et al., 2005) concluded that hippopotamids are a sister group of the anthracotheres in the Hippopotamoidea, instead being suoid, with the Hippopotamoidea being the sister group of cetaceans. This is the *anthracothere hypothesis* (for the origin of hippos). Studies of Boisserie and Fabrice Lihoreau have suggested a bothriodontine *Brachyodus*-like ancestor for the hippopotamids, and Lihoreau and colleagues (2015) have identified *Epirigenys* from the late Oligocene Lokone site as the closest sister group of the hippopotamids. Hippopotamids first appeared by about 21 Ma with *Morotochoerus ugandensis* from the early Miocene of Moroto in Uganda (Orliac et al., 2010). If Boisserie et al. (2005) and Lihoreau et al. (2015) are indeed correct, the hippos originated from an ancient African Oligocene lineage of anthracotheres. In other words, the hippos originated and evolved autochthonously in Island Africa. It should be noted, however, that the current phylogenies in support of the anthracothere hypothesis need further testing on the basis of new fossil discoveries, particularly for the key late Oligocene period, still poorly known in Africa.

CHAPTER 6

An Overview of Mammal Evolution in Island Africa

During the Cenozoic, mammals evolved in the isolated context of an island continent for 43 Ma, from 66 to 23 Ma. The long period of endemic evolution in Island Africa includes four critical times of global changes and evolution that affected ancient African faunas and ecosystems: (1) the Cretaceous–Paleogene (KPg) transition (66 Ma), (2) the Paleocene–Eocene transition (PET, 56 Ma), (3) the Eocene–Oligocene transition (EOT, 34 Ma), and (4) the Oligocene–Miocene transition (OMT, 23 Ma). These global events and their biotic impact are well known in Laurasian continents, but much less well understood in Gondwanan continents, in particular in Island Africa.

Cretaceous–Paleogene transition, 66 Ma

Evolution at the KPg transition, which is a key period of the modernization of the faunas, remains entirely unknown in Africa because of the lack of any known latest Cretaceous or early Paleocene mammal fossils. We can only make a few inferences based on the earliest-known fossils, dated from 60 to 50 Ma (middle–late Paleocene and early Eocene) and based on the phylogeny of the African groups. During at least the Paleocene, Africa was the home of the initial evolutionary radiation of placentals with the evolution of the major clade Afrotheria and probably also of the strepsirrhine primates. Other mammals that evolved at this time in Africa, according to known fossil discoveries, were the hyaenodonts, some extinct basal primates, and stem placentals. No known archaic Mesozoic mammals (mammaliaforms and nontherian mammals) have been discovered in this period or in the Cenozoic of Africa. In particular, multituberculates are still unknown in the Late Cretaceous and Cenozoic of Africa in contrast to Laurasia. Major extinction events occurred in Africa during the Cretaceous–Paleogene transition and in the Late Cretaceous (Rage and Gheerbrant, 2020), perhaps even more dramatically than elsewhere because of the ancient endemism of African faunas that were isolated since the mid-Cretaceous, at around 110–100 Ma. These extinction events decimated the ancient autochthonous fauna and left the African niches free for the evolution of new—that is, immigrant—lineages, such as the placental mammals. The KPg extinction crisis favored the establishment and radiation of placentals in Africa.

Paleocene–Eocene transition, 56 Ma

The Paleocene–Eocene transition at 56 Ma was marked by a global climatic warming event called the Paleocene–Eocene Thermal Maximum (PETM), during which temperatures increased throughout the world by 5°C–8°C. Important faunal changes are recorded at this time, including the extinction of basal placental lineages such as "condylarths" and the appearance of major modern placental orders, such as euprimates, artiodactyls, and perissodactyls, as well as true carnivores (Gingerich, 2006). This period was also a time of great intercontinental dispersals (geodispersals), particularly between Asia, Europe, and North America, which redrew the paleobiogeographic distribution of mammals.

In Africa the evolution of mammals during this time was documented by only two main localities in Morocco (Ouled Abdoun and Ouarzazate Basins) and one in Algeria (El Kohol). One apparent event documented by the Moroccan sites is the extinction of the stem paenungulates, such as ocepeiids, leaving a place for the diversification of modern (extant) orders of paenungulates, such as proboscideans, hyracoids, and embrithopods, which rapidly evolved during the Eocene. Proboscideans show the earliest trend in the evolution of lophodont teeth at the beginning of the Eocene, corresponding to the early occupation of the African folivorous dietary niche, and an increase in body size from small hyrax-sized species (*Eritherium azzouzorum*) to dog-sized species (*Phosphatherium escuilliei*). Trends in increase of body size accelerated during the early Eocene, especially during the so-called Early Eocene Climatic Optimum (EECO) event, which is dated to 53–51 Ma (middle Ypresian). Evolution during this period led to the appearance of the first large African mammals, such as the early proboscideans *Daouitherium* (about 200 kg) and *Numidotherium* (about 250 kg; shoulder height of 1 m.). Other mammal groups document rather steady and continuous evolution across the Paleocene–Eocene transition. This is the case with the small insectivore-like eutherians and placentals (adapisoriculids, palaeoryctids, todralestids) and the first hyaenodonts belonging to koholiine lineage. Such continuity in the evolution of the small insectivore-like eutherians and placentals is also observed in Europe between the Cernay and Dormaal levels, including for the adapisoriculids also known in Africa. If we discard palaeoryctid-like species, a tooth from the Ouarzazate Basin sites (**see fig. 4.2**) indicates that the tenrecoid zalambdodont molar pattern likely developed earlier than the Eocene.

Perhaps the most remarkable event that marks the beginning of the Eocene and the PETM in Africa is the arrival of rodents, represented by anomaluroids. Anomaluroids are currently only known from the early–middle Eocene (about 48 Ma), but their relationships and the evolutionary stage of their earliest representatives (zegdoumyids) suggest they evolved earlier in Africa, likely at the beginning of the Eocene (if not earlier).

Overall, faunal evolution at the Paleocene–Eocene transition seems to be more continuous and conservative in Island Africa than in Laurasian continents with no major faunal turnover, although we should bear in mind the enormous fossil gaps in the period of the Paleocene–Eocene transition in Africa.

Eocene–Oligocene transition, 34 Ma

The Eocene–Oligocene transition was a period of global faunal changes known as the Grande Coupure in Europe. It is the best-documented transitional period of the Paleogene in Africa, thanks especially to the rich Fayum localities. The Grande Coupure faunal turnover is linked to global cooling marked by the first development of Antarctic ice, which caused drought in continents and a major drop in sea levels allowing new intercontinental connections. The Eocene–Oligocene event led to major extinctions of mammals in Laurasia, such as those of the adapiform and omomyiform euprimates, and to wide geodispersals and new radiations of modern taxa (mostly originating from Asia) between northern continents.

In Africa, there were faunal changes at the Eocene–Oligocene transition, but it is far from a faunal break when compared to the Grande Coupure in Europe and other northern continents. The most noticeable faunal changes in Africa from this time correspond to an acceleration of the evolution of endemic groups. African extinctions are limited with respect to the extinction crisis known in Laurasia. A dozen African genera, among primates, rodents, and chiropterans, went extinct, but no major lineage was eliminated. In the case of the Fayum, the faunal changes were progressive, and they extended across the entirety of the Jebel Qatrani Formation—that is, during almost the entire duration of the Rupelian stage from about 34 Ma to 29 Ma. They involved variations in systematic composition and differences in diversity and relative abundance of the groups, especially among the hyracoids and primates (Gagnon, 1997; Rasmussen and Simons, 1991, 1992). For example, the abundance of hyracoids decreased while that of anthracotheres increased. These changes through the Jebel Qatrani Formation were used by Rasmussen and colleagues (1992) to define four faunal zones (FFZ 1, 2, 3, and 4). For Gagnon (1997), these faunal changes were related to local paleoecological modifications in forest biotopes.

A recent large-scale analysis of African faunas concluded that, between 34 and 30 Ma, there was a major decrease in diversity of small mammals, such as primates, rodents, and hyaenodonts, with the extinction of up to two-thirds of the species (de Vries et al., 2021). This is especially noticeable with respect to the Fayum-L41 locality, which is very rich, having yielded half of all the species (35 species) found in other localities of the Jebel Qatrani Formation. The decrease in faunal diversity reported by de Vries and colleagues (2021) is consistent with a lengthy trend of extinctions during the early Oligocene (with at least two peaks of loss of diversity), between 34 and 30 Ma, rather than a single major extinction event at the Eocene–Oligocene transition like the Grande Coupure. This was related to long-term environmental changes in Africa during the early Oligocene.

Africa contrasted with Laurasia by maintaining relatively stable, warm, wet, and tropical climates across the Eocene–Oligocene boundary (Rasmussen et al., 1992). Although African evolution was certainly affected at this time, the tropical low-latitude location of the continent and its isolation dampened the impact of environmental (global cooling) and paleobiogeographic (geodispersal event) changes that occurred in Africa at the beginning of the Oligocene.

The main feature of African mammal evolution at the Eocene–Oligocene transition was its continuous nature. The lineages that continued to evolve are much more significant than reported extinctions (which are highly sensitive to gaps in the fossil record) at that time in Africa. All major groups, such as hyracoids, rodents, strepsirrhine primates, and hyaenodonts, crossed the Eocene–Oligocene boundary—but so did many taxa of lower rank, such as most late Eocene mammal families and several genera of hyaenodonts (*Masrasector, Apterodon*), hyracoids (*Antilohyrax, Bunohyrax*), proboscideans (*Barytherium, Moeritherium*), embrithopods (*Arsinoitherium*), and primates (*Afrotarsius*). This is associated with a small number of early Oligocene taxa (herpetotheriid marsupials, pangolins, and cricetid rodents) that immigrated into Africa.

In fact, the Eocene–Oligocene transition is at the heart of the endemic period of evolution in Africa. The early Oligocene is marked by a rapid evolution of the endemic lineages, with the emergence of major new modern groups, such as elephantiform proboscideans and catarrhine primates, and the radiation and advanced specializations of other lineages, such as hyracoids, embrithopods, rodents, hyaenodonts, and ptolemaiids. This was the time of the evolutionary explosion of hyraxes, anthropoids, rodents, and hyaenodont predators. This was also the time of the evolution of the most spectacular or strange endemic groups, such as the ptolemaiids and the giant arsinoitheres.

Oligocene–Miocene transition, 23 Ma, and establishment of the modern African fauna

At the Oligocene–Miocene transition, around 23 Ma, a major geodynamic event occurred: the closing of the Tethys Sea between Africa and Eurasia, which followed the collision and connection of these two continents **(fig. 6.1)**. This time marks the end of the history of Island Africa, and the birth of the current Old World biogeographic realm that unites Africa, Europe, Asia, and India. The consequence was a faunal revolution in the African continent, the most important in its Cenozoic history. The intercontinental connection of Africa led to a major geodispersal event of mammals; in particular, it led to a massive immigration of Eurasian mammals, which radically changed the composition and evolution of the African fauna. This was the Great Old World Interchange (GOWI; Rage and Gheerbrant, 2020). The colonization of Africa did not happen all at once, but rather in several episodes of dispersals through the new intercontinental land connections established in the Middle East region, across the "*Gomphotherium* land bridge"—so named in allusion to the early emigration of the proboscidean *Gomphotherium* from Africa.

The first-known dispersals from Eurasia occurred between 22 and 18 Ma. In fact, several episodes of faunal dispersals in Africa followed one another, according to the repeated extensions and retractions of the epicontinental seas in the Mediterranean and the Persian Gulf **(fig. 6.1A)**. Paleontologists such as Van der Made (1999) have counted at least eight Miocene faunal dispersal events between Africa and Eurasia up to the Messinian, around 6 Ma. The connection between Africa and Eurasia leads at the same time, by symmetry, to out-of-Africa dispersals of endemic lineages as far as India and beyond to the far east of Asia. The first

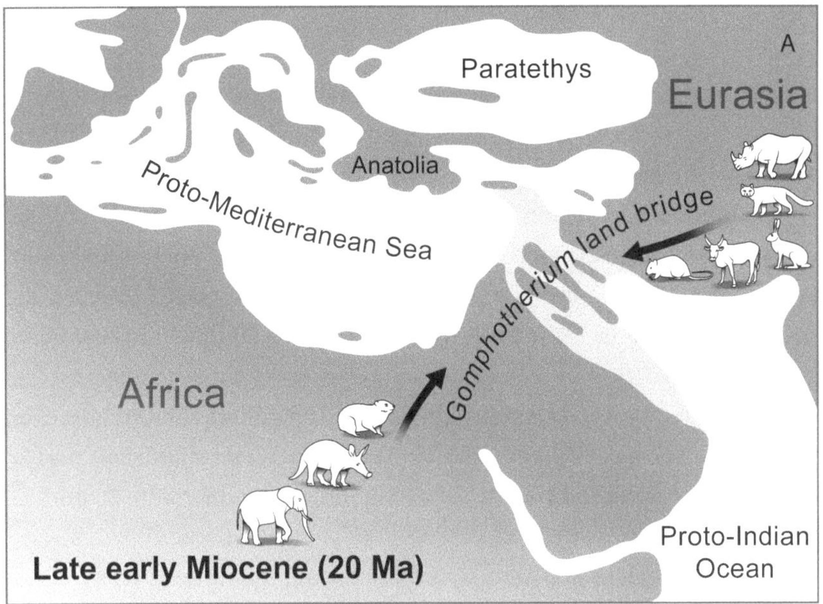

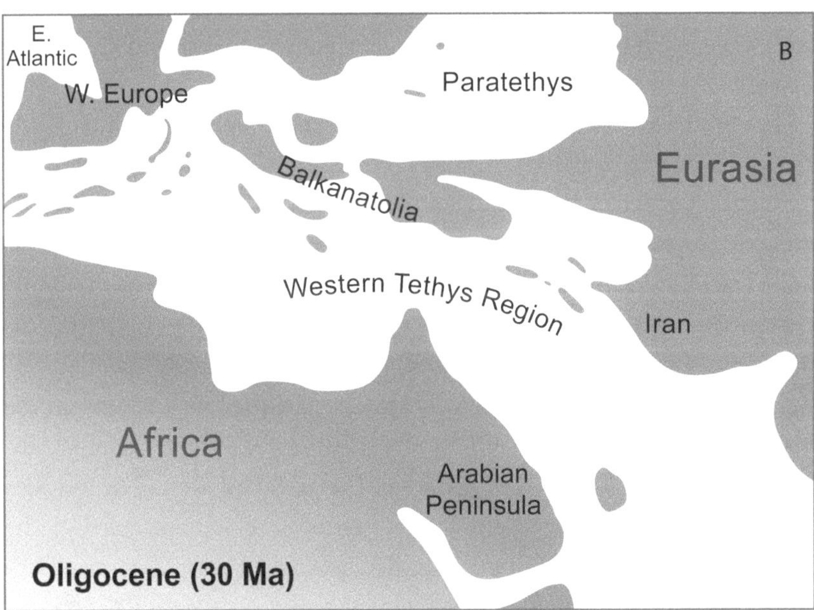

Figure 6.1. The end of Island Africa and the birth of the Old World realm. Paleogeographic evolution of the contact and faunal interchange zone between Africa and Eurasia from the Oligocene (**B**) to the Miocene (**A**) and the emergence of the so-called *Gomphotherium* land bridge. This land bridge was a major early Miocene dispersal corridor established between Africa and Eurasia, in the present-day Middle East. It enabled massive intercontinental faunal interchanges, a geodispersal event between 23 and 18 Ma. These interchanges included important out-of-Africa dispersals of African mammals into Eurasia, such as elephants, hyraxes, aardvarks, and anthropoid primates, and dispersals into Africa of an even greater number of Eurasian mammals, such as modern carnivores (felines, hyaenas, bears, dogs, mongooses), horses, rhinos, giraffes, pigs, shrews, hedgehogs, squirrels, gerbils, pikas, and rabbits. Redrawn by A. Lethiers from Rögl (1998), Harzhauser et al. (2007), and Sen (2013).

episode of the colonization of Eurasia from Africa was marked by the dispersal of elephants (proboscideans), among other mammals, around 20–18 Ma (at the MN3 reference level, "MN" meaning Mammal Neogene). It is called the "Proboscidean Datum Event."

The Eurasian mammal colonization of Africa led to competition with the ancient African natives, resulting in many extinctions. However, the decline of the native African groups was partial and progressive. Some groups, such as the embrithopods, palaeomastodonts, parapithecids, and the stem catarrhines, disappeared sooner, while others resisted and even flourished. The survival and evolutionary success of a significant part of the ancient African natives in modern fauna is a remarkable feature of African evolution. Better still, some of the ancient African lineages, such as the proboscideans and anthropoid primates (monkeys and apes), experienced an evolutionary boom. Nearly all the afrotherians survived (albeit with varying degrees of success), except for the embrithopods, which disappeared.

The modern elephants (elephantoids) exploded in diversity during the Neogene, and they colonized all continents with the exception of the continental islands of Antarctica and Australia. The extant family of elephants (Elephantidae) originated during this diversification. The large pliohyracid hyracoids diversified, and the procaviid family of the living dassies appeared at the beginning of the Miocene and experienced a minor evolutionary radiation. The first appearance of the aardvarks (tubulidentates) in the Miocene is a false appearance due to a huge gap in the fossil record: The relationship of aardvarks to other afrotherians indicates a long ghost lineage dating back to the early Cenozoic. The modern elephant shrews (extant macroscelidid family), which appeared in the site of Eocliff, diversified into three lineages during the Miocene. They comprised the two extant subfamilies Macroscelidinae and Rhynchocyoninae. African phiomorph rodents became abundant in the Miocene, with the appearance and diversification of the modern families of mole-rats (Bathyergidae), dassie-rats (Petromuridae), and cane rats (Thryonomyidae). The Old World monkeys and the hominoids (apes and humans) underwent a remarkable evolutionary radiation, from which the human family is derived.

The modern African fauna owes its establishment and composition to this early Miocene faunal event. Mammals found today in Africa are a mixture of ancient natives (but with more or less ancient lineages) and recent Laurasian groups that arrived in quantity during the Miocene, and the evolution of which (success and extinction) was shaped by recent climatic events.

The African homogeneity in time and space

Current paleontological knowledge—taking into account the many fossil gaps—depicts an overall stable faunal evolution of the modern mammals in Island Africa during the early Cenozoic, from 66 to 23 Ma. The African endemic fauna has been less affected by biotic crises and has evolved more continuously than in Laurasian continents. Major changes in the evolutionary history of African faunas are related both to rapid radiations of endemic lineages and to faunal additions from other continents (i.e., dispersal events), rather than to extinction crises. This is fairly well documented for the early Cenozoic. For earlier periods, in the Mesozoic, the fossils are rare or unknown (e.g., KPg transition). However, the few known vertebrate fossils from the Cretaceous suggest that the ancient native African fauna inherited from Gondwana suffered from important extinctions during the middle Cretaceous with the Cretaceous Terrestrial Revolution, between 125 and 80 Ma, and at the KPg transition (66 Ma) (Rage and Gheerbrant, 2020).

The fauna from Island Africa also displays an overall geographic homogeneity. During the Paleogene period it is very similar across the known African sites, from east to west and from north to south. In other words, no strong provincialism is known in Island Africa. The composition of the mammalian fauna in terms of orders, families, and even genera is globally similar in the sites of the same time span. For example, the early Oligocene localities in Angola have yielded a mammalian association close to that from the Fayum, in Egypt, with the presence of the same genera, and even two similar species. This certainly does not mean that a provincialism did not exist at all in Island Africa. Some southern African faunas, such as the Oligocene Namibian sites (e.g., Eocliff), have, for instance, a peculiar composition with a high diversity of tenrecoids that are much more specialized and closer to modern families than those known in the Fayum faunas. A southern African provincialism has also been reported for the Eocene flora. There was certainly a biogeographic zonation in Island Africa faunas and floras in relation to the climatic zonation of the continent (e.g., amount of rainfall), given its immense latitudinal extent (8,000 km from north to south). Nevertheless, the most remarkable feature of the Island Africa fauna is its endemic character and its general similarities across the continent.

In sum, faunal evolution in Island Africa has been generally continuous (although with periods of more rapid evolution) and conservative. This is related to the endemic context (isolation) of Island Africa and to more stable environmental conditions than in the northern (Laurasian) continents, especially the climates due to the lower latitudinal position of Africa.

The success of the African story: An Island Africa connected

A secondary cradle of placental evolution in Africa

The isolated geographic context of the Arabian-African continent during the Paleogene (i.e., Island Africa) led to the evolution of a remarkable fauna of original mammals. It favored the evolution of one of the three major placental clades, the Afrotheria, and it is at the origin of many modern families that are part of the living African fauna. Island Africa was a major continental center of endemism—indeed, one of the main centers of origin and diversification of the modern mammals, not to mention other vertebrates.

Island Africa was the cradle of an important quantity of extant mammal lineages: one-third of the 18 known placental orders, 5 other suprafamilial clades, and 13 families of afrotherians, primates, and rodents **(table 6.1)**. It was also at the origin of several remarkable extinct groups

Table 6.1.

Main placental mammal groups originating in Island Africa

(* extant; † extinct; + present; − absent)

Taxa			Island Africa origin	Out of Island Africa radiation	Old World radiation (Neogene)
Afrotheria*			+	+	+
	Tenrecoidea*		+	−	−
		Chrysochloridae*	+	−	−
		Potamogalinae*	+	−	−
		Tenrecinae*	+	Malagasy	−
	Macroscelidea*		+	−	+
	Tubulidentata*		+	Malagasy?	+
	Hyracoidea*		+	−	+
	Embrithopoda†		+	N. Tethys, Balkanatolia	−
	Sirenia*		+	Cosmopolitan	Cosmopolitan
	Proboscidea*		+	−	+
Ptolemaiida†			+	−	−
	Ptolemaiidae†		+	−	−
Hyaenodonta†			+	Laurasia	
	Koholiinae†		+	−	−
	Hyainailouroidea†		+	Laurasia (Europe)	−
		Teratodontinae†	+	−	−
		Hyainailouridae†	+	Europe (Apterodontinae)	−
Primates Strepsirrhini*					
	Lorisiformes*		+	−	+
		Lorisidae*	+	−	+
		Galagidae*	+	−	−
	Lemuriformes*		+	Malagasy	−
	Chiromyiformes*		+	Malagasy	−
Primates Anthropoidea*					
	Parapithecoidea		+	South America	−
	Platyrrhini		+	South America	−
	Catarrhini*		+	−	+
	Cercopithecoidea*		+	−	+
	Hominoidea*		+	−	+
Rodentia Hystricognathi*			+	+	−
	Phiomorpha*		+	−	−
		Thryonomyidae*	+	−	−
		Petromuridae*	+	−	−
		Bathyergidae*	+	−	−
	Caviomorpha*		+	South America	−

(continued)

Table 6.1.

Main placental mammal groups originating in Island Africa

(* extant; † extinct; + present; − absent)

Taxa		Island Africa origin	Out of Island Africa radiation	Old World radiation (Neogene)
Rodentia Anomaluroidea*		+	−	−
	Anomaluridae*	+	−	−
	Zenkerellidae*	+	−	−
	Pedetidae*	+	−	−
Carnivora*				
	Stem group of Eupleridae*	+	Malagasy (Eupleridae*)	−
Chiroptera*				
Noctilionoidea*				
	Myzopodidae*	+	Malagasy	−
Artiodactyla*				
Hippopotamoidea*		+	−	+
	Hippopotamidae*	+	−	+

of mammals, such as the hyaenodonts (especially the Hyainailouroidea), the ptolemaiids, and the embrithopods.

The evolution of these endemic mammals was quite successful in the isolated context of Island Africa with little evolutionary competition. Groups such as the hyracoids underwent a spectacular adaptive radiation that saw the evolution of numerous lineages, with diets and locomotion as varied as those found in both the perissodactyls and artiodactyls (euungulates). The sea cows (sirenians) and the elephants (proboscideans) spread widely beyond Africa throughout much of the world during the Oligocene and Neogene. A group of endemic carnivorous mammals, the hyainailouroids, also diversified widely. Some of them specialized with peculiar adaptations, such as the semiaquatic pterodontines that ate shelled prey such as crabs and shrimps. A lineage of hyainailouroids evolved in a particular niche into amazing giant predators (e.g., *Megistotherium*) during the Miocene. Other remarkable African endemic radiations are those of the parapithecoids stem anthropoids and of the anomaluroid (two extant families) and phiomorph (three extant families) rodents.

Island Africa was also the cradle of some major extant mammal lineages that dispersed and diversified out of Africa. These include the South American caviomorph rodents (guinea pigs, chinchillas, agoutis), the South American monkeys (platyrrhines), and the Malagasy tenrecs (tenrecines), lemurs, and carnivorans.

The evolutionary success up to the present day of the ancient native mammals, as part of the living fauna, is an important feature of African evolution. For instance, the native ungulate-like mammals—the hyracoids, elephants, and sea cows—are remarkable components of the extant fauna. They survived to the GOWI following the closure of the Tethys. This contrasts, for example, with the evolution of the South American native ungulate fauna (astrapotheres, litopterns, notoungulates). These ungulates did not survive the end of South American isolation and the Great American Biotic Interchange that followed the connection with North America at 2.5 Ma.

Another notable feature of the fauna of Island Africa is that despite its strong endemic character, it was not homogeneous in its composition and evolution over time. This is an important difference from other insular continents, such as Australia or South America. The fauna of Island Africa is composite in the sense that it comprises groups of different geographic and temporal origins **(fig. 6.2)**. It evolved from various Laurasian stem groups that colonized Island Africa at different times during the Late Cretaceous and the Paleogene. For example, primates of the monkey, ape, and human group (the anthropoids) and important lineages of extant rodents (phiomorphs) arrived in Africa from Asia only in the middle–late Eocene. Mammalian paleontology and phylogeny indicate that most of the native African mammals are related to allochthonous stem groups: Placental mammals that evolved in Island Africa originated from a Laurasian colonization. Favorable local African environments and ecosystems with available food resources and low competitive pressure have stimulated the settlement and evolution of various Eurasian immigrant mammal lineages that arrived in Africa at different times during the Cretaceous and Paleogene **(fig. 6.2)**. It led to spectacular endemic adaptive radiations such as those of afrotherians, catarrhine primates, hyaenodonts

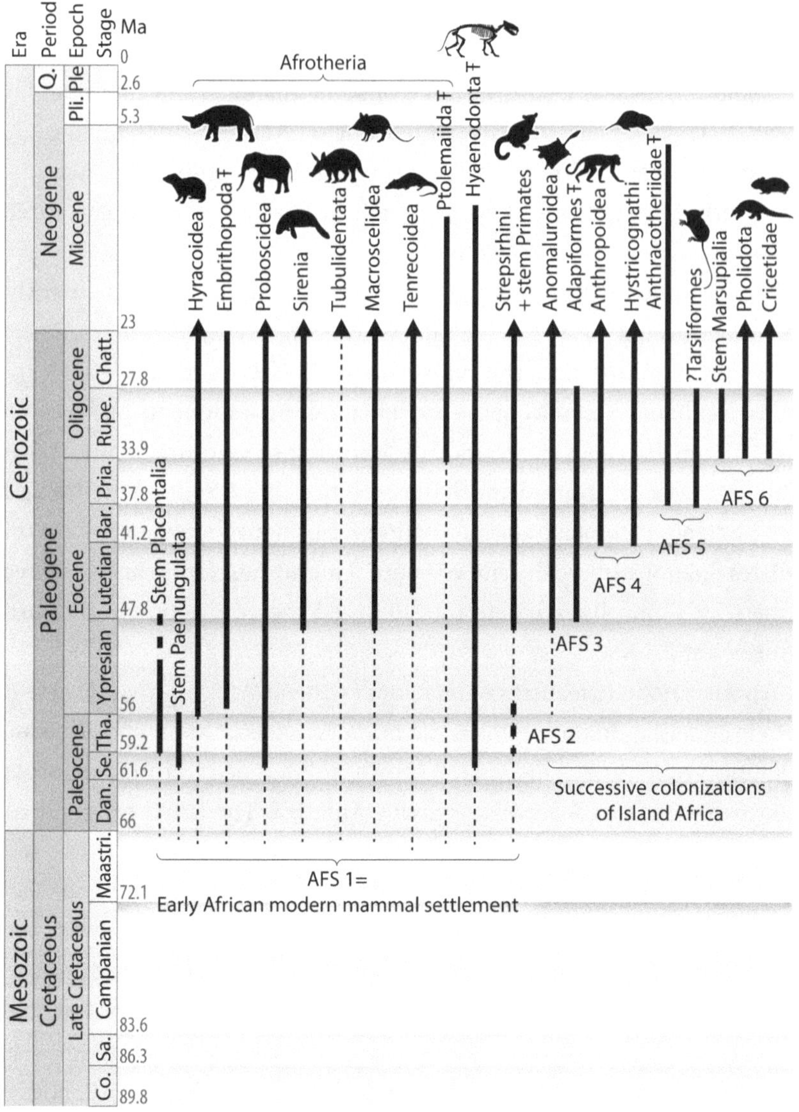

Figure 6.2. Stratigraphic extents of the major lineages of modern mammals (placentals and marsupials) known in Island Africa and the successive episodes of mammalian colonization of the continent. AFS 1–6: Successive African faunal strata (AFS) resulting from several dispersals events (immigrations) from Eurasia across the Tethys Sea. It should be noted that, according to current consensus, the extant hippos originated from an Oligocene African bothriodontine anthracothere (anthracothere hypothesis); this means that anthracotheres are not an entirely extinct lineage. The possible dispersal of the Tarsiiformes in Africa during the late Eocene (AFS 5) is based on the Afrotarsiidae from the Fayum and Dur at Talah, the relationships of which remain disputed (see chapter 6, "A Multiphase African Settlement"). The *solid bar* represents the known stratigraphic fossil record for lineages (clades). The *broken line* represents possible stratigraphic extension based on phylogenetic relationships for some lineages (i.e., ghost lineages). The *broken bar* indicates uncertain relationship and systematic position for some described species (e.g., *Altiatlasius koulchii* from Adrar Mgorn, **fig. 4.3).** Extant lineages are indicated by a stratigraphic extension *bar tipped with an arrow*, and Ŧ indicates extinct lineage. Graph produced by A. Lethiers.

(hyainailouroids), and anomaluroid and phiomorph rodents. In this respect, Island Africa acted as a *secondary cradle* for placentals (Rage and Gheerbrant, 2020). This is also true for other vertebrates.

The big question remaining is, When did placentals first arrive and evolve in Africa? It was well before the first-known fossils dated at 60 Ma (middle Paleocene), but by how much time? Molecular phylogeny suggests that Afrotheria originated in the Late Cretaceous, although no fossil of this age in the world has been yet identified as true placental. The presence of stem placentals such as cimolestid-like eutherians in the Paleocene of Morocco (Gheerbrant, 1992) is consistent with ancient episodes of mammal dispersals from Laurasia, at least as early as the KPg transition.

A multiphase African settlement

Paleontology indicates that the extant and ancient endemic African fauna is derived from Laurasian stem groups that arrived at different times in Island Africa, in several successive waves of colonization. Several dispersals of mammals and other vertebrates, via intermittent trans-Tethyan terrestrial routes and on plant rafts (floating islands), occurred between Laurasia and Africa during the Cretaceous and Paleogene. These dispersals enabled the immigration of various eutherians and placentals, which settled and evolved, some more successfully than others, in Africa **(fig. 6.2)**.

The Cretaceous–Paleogene colonization of Africa by Eurasian placentals resulted in the evolution of several successive African faunal assemblages that are called "African faunal strata," following G. G. Simpson's terminology (Gheerbrant and Rage, 2006). At least six distinct African faunal strata of Paleogene mammals are identified (Gheerbrant and Rage, 2006; Rage and Gheerbrant, 2020). They successively added to each other and evolved during the Paleogene in Island Africa **(fig. 6.2)**. Each African faunal stratum includes Laurasian immigrants (i.e., the stem groups) and all related derived lineages that subsequently evolved as endemics in Africa. The extant native African mammals are derived from the evolution of the lineages that comprise these African faunal strata.

The exact age of recognized successive African faunal strata is reasonably well established for some, less so for other. The earliest-known

African faunal stratum includes the afrotherians, some basal primates (*Altiatlasius*), and probably the strepsirrhines, the hyaenodonts, and several stem placentals, such as the insectivore-like mammals found in the Paleocene of Morocco (e.g., *Cimolestes, Palaeoryctes, Todralestes, Afrodon*). Its age dates to at least the KPg boundary, at 66 Ma, and probably earlier. It is one of the most important components of the African placental fauna, being at the origin of the extant elephant shrews, golden moles, potamogales, tenrecs, aardvarks, dassies, sea cows, elephants, galagos, and lemurs. The major endemic radiation of the hyracoids in Island Africa belongs to this early African faunal stratum. The earliest African rodents represented by stem anomalures (anomaluroids) arrived later as part of the Thanetian–Ypresian African faunal stratum, by 56 Ma. The extinct adapiform primates, which arrived in Africa after the anomalure rodents, belong to an African faunal stratum of at least early–middle Eocene age, around 48 Ma. A more recent middle Eocene African faunal stratum includes the phiomorph rodents (hystricognaths) and the anthropoid primates that appeared at the same time in Africa during the Bartonian. It is of late middle Eocene age (Lutetian–Bartonian boundary), about 41 Ma, and of Asian origin. It gave rise to the monkeys, apes, and humans in Africa. Anthracotheres and possibly tarsiiform primates belong to a late Eocene African faunal stratum dated about 37 Ma. According to the anthracothere hypothesis (see chapter 5), this late Eocene African faunal stratum is at the source of extant hippos. The African herpetotheriid marsupials, the pangolins (Pholidota), and the cricetid rodents belong to a more recent African faunal stratum of earliest Oligocene age, 34 Ma.

The roads across the Tethys Sea to the Promised Land

The colonization of Island Africa by various Laurasian mammals during successive Cretaceous and early Cenozoic dispersal episodes across the Tethys Sea was aided by the peculiar structural and tectonic pattern of the Tethyan Mediterranean area. It is characterized by the presence of the so-called Mediterranean Tethyan Sill (Dercourt et al., 1993; Vrielynck et al., 1994), which consists of a series of continental platforms and microplates—such as the Alboran (part of the Iberian Peninsula), Apulia (part of the

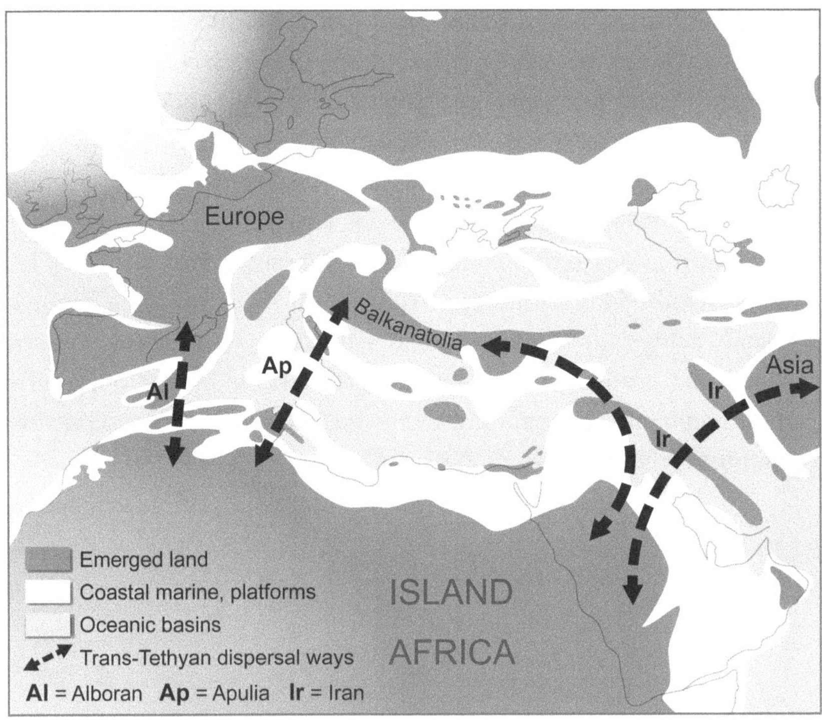

Figure 6.3. Paleogeography of the Tethyan area ("Mediterranean Tethyan Sill"), between Africa and Eurasia, and possible dispersal routes for mammals and other land vertebrates across the Tethys Sea during the Lutetian (45 Ma). Note the intermediate insular archipelago of Balkanatolia, which extended between Africa and Europe as a big stepping stone. Early African mammals, such as embrithopods, have been found in Balkanatolia (e.g., Métais et al., 2024). Redrawn by A. Lethiers from Gheerbrant and Rage (2006, fig. 3).

Italian Peninsula), Taurides (part of Anatolia, Turkey), and Iran—that were emergent during periods of low sea level and acted as intermediate stepping stones for vertebrate dispersals. Some of these continental microplates formed the north Tethyan archipelago called Balkanatolia **(fig. 6.3)** (Licht et al., 2022), in which African mammals, such as embrithopods of the family Palaeoamasiidae, have been found. The Mediterranean Tethyan Sill did not form continuous land bridges between Africa and Eurasia, but rather transitory and discontinuous land routes controlled by sea-level changes during the Cretaceous and Paleogene. It acted as a selective filter for the dispersal of continental vertebrates between

Laurasia and Island Africa (Gheerbrant and Rage, 2006). Passive trans-Tethyan dispersals by rafting ("sweepstake dispersals") also occurred, particularly in times of global warming causing major floods.

The occasional but repeated paleobiogeographic links between Island Africa and Eurasia across the Tethys Sea allowed the faunal colonization of this Gondwanan continental province, especially by placentals after the KPg extinction event (66 Ma). The immigrant mammals, which arrived in Island Africa via the trans-Tethyan routes or by rafting, found a new and ecologically favorable land of evolution. They evolved and diversified successfully into several and remarkable endemic lineages, such as elephants, anomalure rodents, galagos, and anthropoids, that are important components of the present-day fauna.

What Is Next in the Quest for African Origins?

Africa was a gigantic continental island and a major center of evolution for some 80 million years, from the mid-Cretaceous (110–100 Ma) to the end of the Paleogene (23 Ma). This was a long and important period in the evolution of vertebrates. The fossil record yields evidence concerning the evolution of modern mammals—placentals and marsupials—in Island Africa during the second half of its history, from 66 to 23 Ma (with gaps). It turns out that the endemic African evolution spans the first two-thirds of the Age of Mammals. This was a key period of faunal modernization, especially during the KPg (66 Ma) and Paleocene–Eocene (56 Ma) transitions.

The latest paleontological field research in Africa, carried out over 40 years beginning in the mid-1980s, has considerably increased our knowledge. The fossil record of placental mammals in Island Africa has been extended by 20 million years, with new discoveries in periods that were previously unknown or poorly represented, such as the late Oligocene (28–23 Ma), the early Eocene (56–48 Ma), and the Paleocene (recorded between 60 and 56 Ma). The quantities of fossiliferous localities increased. Their geographic distribution, previously mostly North African, has been extended to southern Africa (Namibia, Angola), East Africa (Tanzania, Eritrea, Kenya), and to the Arabian Peninsula. More than 100 new fossil species have been described: This is one-third of all known species in Africa during the Paleogene! The first articulated skeletons of early African mammals have been described from Libya and Tanzania.

There are still many gaps in the fossil record and a paucity of knowledge about the ancient fauna of Island Africa and its evolution. However, the major features of the early evolutionary history of mammals in Africa are becoming better documented and more clearly understood. A key point is that early African placental mammal fauna evolved in an endemic way in the context of a continental island, but at the same time Africa maintained episodic faunal links with faunas from other continents, especially Eurasia, and more marginally South America and Madagascar. This contrasts with the "Splendid Isolation" of South America during the Cenozoic.

This paleobiogeographic context shaped the evolution of African faunas. It explains the remarkable endemic adaptive radiation of various original groups that were ultimately rooted in exotic ancestors. Africa was thus a secondary cradle of evolution for several extinct and extant lineages: It acted as an evolutionary incubator fed by diverse ancestral Eurasian mammals. It is noteworthy that most mammalian dispersals into Island Africa came from Eurasia, and that no definite immigration from Gondwanan continents (e.g., South America) is known. The successful evolution of placental mammals in the African island was boosted by favorable environmental and ecological contexts with low competition pressure for resources. This is due both to the insular context and to probable major extinctions of the ancient native fauna inherited from Gondwana during the Cretaceous, at the time of the Cretaceous Terrestrial Revolution (125–80 Ma) and the KPg (66 Ma) crisis, which opened new ecospaces for mammal evolution in Island Africa.

The intermittent faunal interchanges of Island Africa and Eurasia, facilitated by dispersal routes across the Tethys Sea and by passive transport on plant rafts during monsoon periods, preceded and foreshadowed the great Old World faunal province that was established 23 Ma. The Eurasian dispersals into Island Africa, repeated throughout its history, appear to have stimulated a competitive dynamism, both adaptively and evolutionarily, of the African mammal lineages (Rage and Gheerbrant, 2020). This probably aided the endemic African fauna to survive the wave of dispersals from Eurasia during the Great Old World Interchange that followed the Africa-Eurasia collision 23 Ma. This included some of the most amazing wildlife of the world today, such as elephants and apes, both of which subsequently expanded successfully out of Africa.

A number of important questions about evolution in Africa remain unanswered. One series of questions concerns the origins of the placentals that populated Africa. When did the first placentals reach Island Africa, where did they come from, and what were the early immigrant groups? This touches on the origin of a major group of placentals, the Afrotheria, and even more broadly on the initial diversification of placentals insofar as the Afrotheria is one of the first major branches to have diverged in the evolutionary history of placentals. For instance, a hypothesis of molecular phylogeny is that the endemic African and South American mammals share a common Cretaceous ancestor from a proposed clade Atlantogenata which includes both the Afrotheria and Xenarthra (sloths, anteaters, armadillos). This hypothesis involves unresolved scenarios of dispersals between South America and Africa (the alternative vicariant hypothesis—which implies fragmentation of the geographic ranges of ancestral mammals by continental drifting—being less likely in terms of the age of the divergence of the groups). Competing phylogenetic hypotheses to an atlantogenatan ancestor indicate that afrotherians were either the first or the second branch of placentals to separate during their evolution. In all these hypotheses, the known fossil record (e.g., stem placentals) favors a Laurasian, and especially an Asian, origin of the placentals, and thus an early Laurasian origin of the Afrotheria.

The challenge to elucidating the origin of the first African placentals and the afrotherians is in finding fossils of mammals in Island Africa in the blank period of the Late Cretaceous and the Paleocene, especially between 90 and 60 Ma. A further enigmatic aspect of African mammal paleontology concerns the origin and ancient history of the aardvarks (tubulidentates). Their fossil record is a blank prior to 21 Ma, while paradoxically they belong to an ancient afrotherian lineage, possibly as old as the Late Cretaceous.

Another outstanding question is that of the origin and early radiation of the primates in Africa. This is especially true for the lemur, loris, and galago group (strepsirrhines), the earliest-known fossils of which are of middle Eocene age (ca 48 Ma). Similarly, the first African rodents (anomaluroids) are found in the same middle Eocene sites, while their phylogeny implies an early Eocene or Paleocene origin. Also, no fossils of springhares, which are the sister group of the 48-million-year-old anomaluroids, have been found before 23 Ma (basal Miocene). The same question arises for the origin of the Malagasy carnivores, which, according to molecular phylogeny, are likely to have their origins in the African Oligocene, by 25 Ma, although there is still no paleontological evidence for this.

Finally, available paleontological evidence regarding mammals confirms that Africa acted as a great evolutionary laboratory for numerous lineages, and most spectacularly when it was a continental island. The fossil record documents the role of this natural laboratory over a period of 60 million years in the emergence of the extant mammals, including the primates at the root of the human family, among many other notable groups.

Despite spectacular progress, more paleontological field research, including prospecting for fossils and excavations, is essential for yielding new evidence regarding paleobiodiversity and evolution in Africa. The immense "lost world" of Island Africa has not yielded all its secrets—there is much more to learn about the remarkable extinct animals and the rich and ancient evolutionary history of the extant fauna.

Appendix

Faunal list of mammal species recorded in the Paleogene sites of Africa. Genera and species are listed in stratigraphic order, from the earliest-known Paleogene sites upward. aff. = *affinis* (open nomenclature); BOTM = bed with Oligocene terrestrial mammals; cf. = *confer* (open nomenclature); CTMD = Continental and Transitional Marine Deposits; Fm = formation; Indet. = indeterminable species; K/Ar = potassium–argon radiometric dating method; n. g. = new genus; n. sp. = new species.

Genus and species Author, year	Suprageneric taxonomy	Locality; stratigraphic level	Epoch; age
Lahimia selloumi Solé and Gheerbrant, 2009	Koholiinae; Hyaenodonta	Ouled Abdoun Basin, Morocco; Bed IIa	middle Paleocene; Selandian
Ocepeia daouiensis Gheerbrant and Sudre, 2001	Ocepeiidae; Paenungulatomorpha	Ouled Abdoun Basin, Morocco; Bed IIa	middle Paleocene; Selandian
Eritherium azzouzorum Gheerbrant, 2009	Proboscidea	Ouled Abdoun Basin, Morocco; Bed IIa	middle Paleocene; Selandian
Abdounodus hamdii Gheerbrant and Sudre, 2001	Paenungulatomorpha	Ouled Abdoun Basin, Morocco; Bed IIa	middle Paleocene; Selandian
Ocepeia grandis Gheerbrant, 2014	Ocepeiidae; Paenungulatomorpha	Ouled Abdoun Basin, Morocco; Bed IIa	middle or late Paleocene; Selandian or Thanetian
Cimolestes cf. *incisus* Marsh, 1889	Didelphodontinae, Palaeoryctidae; Cimolesta	Ouarzazate Basin, Morocco; Jebel Guersif Fm	late Paleocene; Thanetian
Palaeoryctes cf. *minimus* Gheerbrant, 1992	Palaeoryctinae, Palaeoryctidae; Cimolesta	Ouarzazate Basin, Morocco; Jebel Guersif Fm	late Paleocene; Thanetian
Cimolestes cuspulus Gheerbrant, 1992	Didelphodontinae, Palaeoryctidae; Cimolesta	Ouarzazate Basin, Morocco; Jebel Guersif Fm	late Paleocene; Thanetian
Aboletylestes hypselus? Russell, 1964	Didelphodontinae, Palaeoryctidae; Cimolesta	Ouarzazate Basin, Morocco; Jebel Guersif Fm	late Paleocene; Thanetian
Indet. 1	Didelphodontinae, Palaeoryctidae; Cimolesta	Ouarzazate Basin, Morocco; Jebel Guersif Fm	late Paleocene; Thanetian
Indet. 2	Didelphodontinae, Palaeoryctidae; Cimolesta	Ouarzazate Basin, Morocco; Jebel Guersif Fm	late Paleocene; Thanetian
Palaeoryctes minimus Gheerbrant, 1992	Palaeoryctinae, Palaeoryctidae; Cimolesta	Ouarzazate Basin, Morocco; Jebel Guersif Fm	late Paleocene; Thanetian

Aboletylestes robustus Gheerbrant, 1992	Didelphodontinae, Palaeoryctidae; Cimolesta	Ouarzazate Basin, Morocco; Jebel Guersif Fm	late Paleocene; Thanetian
cf. *Todralestes* sp.	Todralestidae; Eutheria	Ouarzazate Basin, Morocco; Jebel Guersif Fm	late Paleocene; Thanetian
Afrodon cf. *chleuhi* Gheerbrant, 1988	Adapisoriculidae; Eutheria	Ouarzazate Basin, Morocco; Jebel Guersif Fm	late Paleocene; Thanetian
Afrodon chleuhi Gheerbrant, 1988	Adapisoriculidae; Eutheria	Ouarzazate Basin, Morocco; Jebel Guersif Fm	late Paleocene; Thanetian
Garatherium? todrae Gheerbrant, 1998	Adapisoriculidae; Eutheria	Ouarzazate Basin, Morocco; Jebel Guersif Fm	late Paleocene; Thanetian
Tinerhodon disputatum Gheerbrant, 1995	Proviverrinae?; Hyaenodonta	Ouarzazate Basin, Morocco; Jebel Guersif Fm	late Paleocene; Thanetian
Indet.	Hyaenodonta?	Ouarzazate Basin, Morocco; Jebel Guersif Fm	late Paleocene; Thanetian
Indet.1	Tenrecoidea?	Ouarzazate Basin, Morocco; Jebel Guersif Fm	late Paleocene; Thanetian
Indet.	Plesiadapiformes or Primates	Ouarzazate Basin, Morocco; Jebel Guersif Fm	late Paleocene; Thanetian
Altiatlasius koulchii Sigé, 1990	Simiiformes?, Haplorhini?; Primates	Ouarzazate Basin, Morocco; Jebel Guersif Fm	late Paleocene; Thanetian
Indet.	Hyaenodonta ("Viverridae? or Miacidae?" in Gheerbrant (1995))	Ouarzazate Basin, Morocco; Jebel Guersif Fm	late Paleocene; Thanetian
Indet. 2	Paenungulata? ("Condylarthra" in Gheerbrant (1995))	Ouarzazate Basin, Morocco; Jebel Guersif Fm	late Paleocene; Thanetian
Indet.	Paenungulata? ("Ungulata?" in Gheerbrant (1995))	Ouarzazate Basin, Morocco; Jebel Guersif Fm	late Paleocene; Thanetian

(*continued*)

Genus and species Author, year	Suprageneric taxonomy	Locality; stratigraphic level	Epoch; age
Indet. 1	Eutheria ("Proteutheria or Lipotyphla" in Gheerbrant (1995))	Ouarzazate Basin, Morocco; Jebel Guersif Fm	late Paleocene; Thanetian
Indet. 2	Eutheria ("Proteutheria or Lipotyphla" in Gheerbrant (1995))	Ouarzazate Basin, Morocco; Jebel Guersif Fm	late Paleocene; Thanetian
Indet. 3	Eutheria ("Proteutheria or Lipotyphla" in Gheerbrant (1995))	Ouarzazate Basin, Morocco; Jebel Guersif Fm	late Paleocene; Thanetian
Indet. (div. ssp.)	Eutheria	Ouarzazate Basin, Morocco; Jebel Guersif Fm	late Paleocene; Thanetian
Todralestes variabilis Gheerbrant, 1991	Todralestidae; Eutheria	Ouarzazate Basin, Morocco; Jebel Guersif Fm	late Paleocene; Thanetian
Afrodon sp.	Adapisoriculidae; Eutheria	Ouarzazate Basin, Morocco; Jebel Guersif Fm	late Paleocene; Thanetian
Indet.	Paenungulata? (described as "cf. Arctocyonidae")	Ouarzazate Basin, Morocco; Jebel Guersif Fm	late Paleocene; Thanetian
Boualitomus marocanensis Gheerbrant, 2006	Koholiinae; Hyaenodonta	Ouled Abdoun Basin, Morocco; Bed I?	early Eocene?; Ypresian?
Indet.	Hyracoidea	Ouled Abdoun Basin, Morocco; Bed 0 or sillons	early Eocene; Ypresian (middle)
Seggeurius n. sp.	Hyracoidea	Ouled Abdoun Basin, Morocco; Bed I, intercalary IIa/I	early Eocene; Ypresian (early)
Phosphatherium escuilliei Gheerbrant et al., 1996	Phosphatheriidae; Proboscidea	Ouled Abdoun Basin, Morocco; Bed I, intercalary IIa/I	early Eocene; Ypresian
Daouitherium rebouli Gheerbrant and Sudre, 2002	Proboscidea	Ouled Abdoun Basin, Morocco: Grand Daoui; Bed 0	early Eocene; Ypresian (middle)

Stylolophus minor Gheerbrant, 2018	Stylolophidae; Embrithopoda	Ouled Abdoun Basin, Morocco; Bed I, intercalary IIa/I	early Eocene; Ypresian (early)
Indet.	Macroscelidea or Paenungulata	Ouled Abdoun Basin, Morocco; Bed I, intercalary IIa/I	early Eocene; Ypresian
Stylolophus major Gheerbrant, 2018	Stylolophidae; Embrithopoda	Ouled Abdoun Basin, Morocco; Bed 0 or sillons	early Eocene; Ypresian (middle)
Indet.	Hyracoidea	Tamaguilelt (= Tamaguilel), Iullemmeden Basin, Mali	early (or middle) Eocene?
Indet.	Numidotheriidae?; Proboscidea	Tamaguilelt (= Tamaguilel), Iullemmeden Basin, Mali	early (or middle) Eocene?
Indet.	Proboscidea?	Tamaguilelt (= Tamaguilel), Iullemmeden Basin, Mali	early (or middle) Eocene?
Indet.	Eutheria Indet.	Tamaguilelt (= Tamaguilel), Iullemmeden Basin, Mali	early (or middle) Eocene?
Garatherium mahboubii Crochet, 1984	Adapisoriculidae?	El Kohol, Brezina, Algeria; El Kohol Fm	early Eocene; Ypresian
Koholia atlasense Crochet, 1988	Koholiinae; Hyaenodonta	El Kohol, Brezina, Algeria; El Kohol Fm	early Eocene; Ypresian
Seggeurius amourensis Crochet, 1986	Hyracoidea	El Kohol, Brezina, Algeria; El Kohol Fm	early Eocene; Ypresian
Numidotherium koholense Mahboubi et al., 1986	Numidotheriidae, Barytherioidea; Proboscidea	El Kohol, Brezina, Algeria; El Kohol Fm	early Eocene; Ypresian
Indet.	Microchiroptera; Eochiroptera, Chiroptera	El Kohol, Brezina, Algeria; El Kohol Fm	early Eocene; Ypresian
Indet. 2	Didelphodontinae, Palaeoryctidae; Cimolesta	Ouarzazate Basin (NTG2), Morocco; Ait Ouarithane Fm (base)	early Eocene; Ypresian
n. sp.	Didelphodontinae, Palaeoryctidae; Cimolesta	Ouarzazate Basin (NTG2), Morocco; Ait Ouarithane Fm (base)	early Eocene; Ypresian

(continued)

Genus and species Author, year	Suprageneric taxonomy	Locality; stratigraphic level	Epoch; age
Todralestes butleri Gheerbrant, 1993	Todralestidae; Eutheria	Ouarzazate Basin (NTG2), Morocco; Ait Ouarithane Fm (base)	early Eocene; Ypresian
Afrodon tagourtensis Gheerbrant, 1993	Adapisoriculidae; Eutheria.	Ouarzazate Basin (NTG2), Morocco; Ait Ouarithane Fm (base)	early Eocene; Ypresian
Khamsaconus bulbosus Sudre et al., 1993	Phosphatheriidae; Proboscidea	Ouarzazate Basin (NTG2), Morocco; Ait Ouarithane Fm (base)	early Eocene; Ypresian
Indet.	Paenungulata? ("Condylarthra")	Ouarzazate Basin (NTG2), Morocco; Ait Ouarithane Fm (base)	early Eocene; Ypresian
Indet. 1–5	Eutheria Indet.	Ouarzazate Basin (NTG2), Morocco; Ait Ouarithane Fm (base)	early Eocene; Ypresian
Pappocetus lugardi Andrews, 1920	Protocetidae, Archaeoceti; Cetacea	Ameki, Nigeria; Ameki Fm	middle Eocene; Lutetian
Indet.	Sirenia or Cetacea	Borot Ramalieh, loc. PAL.EO, Negev, Israel; Horsha Fm	middle Eocene; Lutetian
Libysiren sickenbergi Domning et al., 2017	Protosirenidae; Sirenia	Bu El Haderait, Libya; Wadi Thamat Fm, al Gata Member	middle Eocene; Lutetian
Indet.	Sirenia	Callis, Carcar, and Mogadishu, Somalia	middle Eocene; Lutetian
Kasserinotherium tunisiense Crochet, 1986	Peradectidae?; Metatheria?	Chambi, Tunisia	middle Eocene; Lutetian
Chambilestes foussanensis Gheerbrant and Hartenberger, 1999	Chambilestidae; Eulipotyphla?	Chambi, Tunisia	middle Eocene; Lutetian
Furodon crocheti Solé et al., 2014	Hyainailouroidea; Hyaenodonta	Chambi, Tunisia	middle Eocene; Lutetian
Parvavorodon gheerbranti Solé et al., 2014	Hyainailourinae, Hyainailouridae, Hyainailouroidea, Hyaenodonta	Chambi, Tunisia	middle Eocene; Lutetian
Indet.	Hyracoidea	Chambi, Tunisia	middle Eocene; Lutetian

Megalohyrax? sp.	Geniohyidae; Hyracoidea	Chambi, Tunisia	middle Eocene; Lutetian
Titanohyrax tantulus Court and Hartenberger, 1992	Titanohyracidae; Hyracoidea	Chambi, Tunisia	middle Eocene; Lutetian
Indet.	Sirenia	Chambi, Tunisia	middle Eocene; Lutetian
Zegdoumys sbeitlai Vianey-Liaud et al., 1994	Zegdoumyidae, Anomaluroidea; Rodentia	Chambi, Tunisia	middle Eocene; Lutetian
Djebelemur martinezi Hartenberger and Marandat, 1992	Djebelemuridae, Strepsirrhini; Primates	Chambi, Tunisia	middle Eocene; Lutetian
cf. *Algeripithecus* sp.	Azibiidae; Strepsirrhini; Primates	Chambi, Tunisia	middle Eocene; Lutetian
cf. *Djebelemur* sp.	Djebelemuridae, Strepsirrhini; Primates	Chambi, Tunisia	middle Eocene; Lutetian
Vespertiliavus? aenigma Ravel et al., 2016	Emballonuridae, Emballonuroidea, Microchiroptera; Chiroptera	Chambi, Tunisia	middle Eocene; Lutetian
Khoufechia gunnelli Ravel et al., 2016	Nycteridae, Emballonuroidea, Microchiroptera; Chiroptera	Chambi, Tunisia	middle Eocene; Lutetian
Vespertiliavus kasserinensis Ravel et al., 2016	Emballonuridae, Emballonuroidea, Microchiroptera; Chiroptera	Chambi, Tunisia	middle Eocene; Lutetian
Hipposideros (Pseudorhinolophus) africanum Ravel et al., 2016	Hipposideridae, Rhinolophoidea, Microchiroptera; Chiroptera	Chambi, Tunisia	middle Eocene; Lutetian
Necromantis? fragmentum Ravel et al., 2016	Necromantidae, Rhinolophoidea, Microchiroptera; Chiroptera	Chambi, Tunisia	middle Eocene; Lutetian
Indet.	Nycteridae, Emballonuroidea, Microchiroptera; Chiroptera	Chambi, Tunisia	middle Eocene; Lutetian
Palaeophyllophora? tunisiensis Ravel et al., 2016	Hipposideridae, Rhinolophoidea, Microchiroptera; Chiroptera	Chambi, Tunisia	middle Eocene; Lutetian

(continued)

Genus and species Author, year	Suprageneric taxonomy	Locality; stratigraphic level	Epoch; age
Dizzya exsultans Sigé, 1991	Philisidae, Vespertilionoidea, Microchiroptera; Chiroptera	Chambi, Tunisia	middle Eocene; Lutetian
Indet.	Vespertilionoidea, Microchiroptera; Chiroptera	Chambi, Tunisia	middle Eocene; Lutetian
Chambinycteris pusilli Ravel et al., 2016	Vespertilionoidea, Microchiroptera; Chiroptera	Chambi, Tunisia	middle Eocene; Lutetian
Witwatia sigei Ravel et al., 2012	Philisidae, Vespertilionoidea, Microchiroptera; Chiroptera	Chambi, Tunisia	middle Eocene; Lutetian
Indet.	Placentalia Indet. (Primates Sivaladapinae?, Hyracoidea?)	Chambi, Tunisia	middle Eocene; Lutetian
Eotheroides aegyptiacum (Owen 1875) Trouessart, 1905	Halitheriinae, Dugongidae; Sirenia	Djebel Mokattam (Le Caire), Egypt; Mokattam Fm (Limestone)	middle Eocene; Lutetian
Protosiren fraasi Abel, 1907	Protosirenidae; Sirenia	Djebel Mokattam (Le Caire), Egypt; Mokattam Fm (Limestone)	middle Eocene; Lutetian
Protocetus atavus Fraas, 1904	Protocetidae, Archaeoceti; Cetacea	Djebel Mokattam (Le Caire), Egypt; Mokattam Fm (Limestone)	middle Eocene; Lutetian
Eosiren abeli Sickenberg, 1934	Halitheriinae, Dugongidae; Sirenia	Djebel Mokattam, Egypt; Mokattam Fm (Limestone)	middle Eocene; Lutetian
Phiomicetus anubis Gohar et al., 2021	Protocetidae, Archaeoceti; Cetacea,	Wadi Al-Ruwayan, Fayum, Egypt; Midawara Fm	middle Eocene; Lutetian
Palaeoryctes sp.	Palaeoryctinae, Palaeoryctidae; Cimolesta	Glib Zegdou, Hammada du Draa, Algeria; Glib Zegdou Fm	middle Eocene; Lutetian
spp. Indet.	Eulipotyphla?	Glib Zegdou, Hammada du Draa, Algeria; Glib Zegdou Fm	middle Eocene; Lutetian

Chambius kasserinensis Hartenberger, 1986	Herodotiinae; Macroscelididae, Macroscelidea	Glib Zegdou, Hammada du Draa, Algeria; Glib Zegdou Fm	middle Eocene; Lutetian
Glibemys algeriensis Vianey-Liaud et al., 1994	Zegdoumyidae, Anomaluroidea; Rodentia	Glib Zegdou, Hammada du Draa, Algeria; Glib Zegdou Fm	middle Eocene; Lutetian
Zegdoumys lavocati Vianey-Liaud et al., 1994	Zegdoumyidae, Anomaluroidea; Rodentia	Glib Zegdou, Hammada du Draa, Algeria; Glib Zegdou Fm	middle Eocene; Lutetian
n. g. 1, n. sp. 1	Anomaluroidea; Rodentia	Glib Zegdou, Hammada du Draa, Algeria; Glib Zegdou Fm	middle Eocene; Lutetian
n. g. 2, n. sp. 2	Anomaluroidea; Rodentia	Glib Zegdou, Hammada du Draa, Algeria; Glib Zegdou Fm	middle Eocene; Lutetian
Glibia pentalopha Vianey-Liaud et al., 1994	Zegdoumyidae, Anomaluroidea; Rodentia	Glib Zegdou, Hammada du Draa, Algeria; Glib Zegdou Fm	middle Eocene; Lutetian
Glibia tetralopha Vianey-Liaud et al., 1994	Zegdoumyidae, Anomaluroidea; Rodentia	Glib Zegdou, Hammada du Draa, Algeria; Glib Zegdou Fm	middle Eocene; Lutetian
Lazibemys zegdouensis Marivaux et al., 2011	Zegdoumyidae, Anomaluroidea; Rodentia	Glib Zegdou, Hammada du Draa, Algeria; Glib Zegdou Fm	middle Eocene; Lutetian
Indet.	Djebelemuridae, Strepsirrhini; Primates	Glib Zegdou, Hammada du Draa, Algeria; Glib Zegdou Fm	middle Eocene; Lutetian
Algeripithecus minutus Godinot and Mahboubi, 1992	Djebelemuridae, Strepsirrhini; Primates	Glib Zegdou, Hammada du Draa, Algeria; Glib Zegdou Fm	middle Eocene; Lutetian
Azibius sp.	Djebelemuridae, Strepsirrhini; Primates	Glib Zegdou, Hammada du Draa, Algeria; Glib Zegdou Fm	middle Eocene; Lutetian
Pseudovespertiliavus parva Ravel et al., 2016	Emballonuridae, Emballonuroidea, Microchiroptera; Chiroptera	Glib Zegdou, Hammada du Draa, Algeria; Glib Zegdou Fm	middle Eocene; Lutetian
Titanohyrax mongereaui Sudre, 1979	Titanohyracidae; Hyracoidea	Glib Zegdou, Hammada du Draa, Algeria; Glib Zegdou Fm	middle Eocene; Lutetian

(continued)

Genus and species Author, year	Suprageneric taxonomy	Locality; stratigraphic level	Epoch; age
Furodon crocheti Solé et al., 2014	Hyainailourinae, Hyainailouridae, Hyainailouroidea, Hyaenodonta	Hammada du Draa, Algeria; Gour Lazib, Glib Zegdou Fm	middle Eocene; Lutetian
Parvavorodon gheerbranti Solé et al., 2014	Hyainailourinae, Hyainailouridae, Hyainailouroidea, Hyaenodonta	Hammada du Draa, Algeria; Gour Lazib, Glib Zegdou Fm	middle Eocene; Lutetian
Glibzegdouia tabelbalaensis Crochet et al., 2001	Hyainailouroidea, Hyaenodonta,	Hammada du Draa, Algeria; Gour Lazib, Glib Zegdou Fm	middle Eocene; Lutetian
Chambius? sp.	Herodotiinae, Macroscelididae; Macroscelidea	Hammada du Draa, Algeria; Gour Lazib, Glib Zegdou Fm	middle Eocene; Lutetian
Titanohyrax? sp. 1	Titanohyracidae; Hyracoidea	Hammada du Draa, Algeria; Gour Lazib, Glib Zegdou Fm	middle Eocene; Lutetian
Titanohyrax? sp. 2	Titanohyracidae; Hyracoidea	Hammada du Draa, Algeria; Gour Lazib, Glib Zegdou Fm	middle Eocene; Lutetian
Titanohyrax tantulus Court and Hartenberger, 1992	Titanohyracidae; Hyracoidea	Hammada du Draa, Algeria: Gour Lazib, Glib Zegdou Fm	middle Eocene; Lutetian
Witwatia sigei Ravel et al., 2012	Philisidae, Emballonuroidea, Microchiroptera; Chiroptera	Hammada du Draa, Algeria; Gour Lazib, Glib Zegdou Fm	middle Eocene; Lutetian
Drakonycteris glibzegdouensis Ravel et al., 2016	Emballonuroidea, Microchiroptera; Chiroptera	Hammada du Draa, Algeria; Gour Lazib/Glib Zegdou, Glib Zegdou Fm	middle Eocene; Lutetian
Azibius trerki Sudre, 1975	Azibiidae, Strepsirrhini; Primates	Gour Lazib locus 1, Hammada du Draa, Algeria; Glib Zegdou Fm	middle Eocene; Lutetian
Megalohyrax gevini Sudre, 1979	Geniohyidae; Hyracoidea	Gour Lazib locus 2, Hammada du Draa, Algeria; Glib Zegdou Fm	middle Eocene; Lutetian
Helioseus insolitus Sudre, 1979	Hyracoidea	Gour Lazib locus 2, Hammada du Draa, Algeria; Glib Zegdou Fm	middle Eocene; Lutetian

Microhyrax lavocati Sudre, 1979	Hyracoidea	Gour Lazib locus 2, Hammada du Draa, Algeria; Glib Zegdou Fm	middle Eocene; Lutetian
Bunohyrax or *Megalohyrax* Indet.	Geniohyidae; Hyracoidea	Gour Lazib locus 3, Hammada du Draa, Algeria; Glib Zegdou Fm	middle Eocene; Lutetian
Indet.	Protosirenidae; Sirenia	Kpogamé-Hahotoé, Togo	middle Eocene; Lutetian
Indet.	Dugongidae; Sirenia	Kpogamé-Hahotoé, Togo	middle Eocene; Lutetian
Togocetus traversei Gingerich and Cappetta, 2014	Protocetidae, Archaeoceti; Cetacea	Kpogamé-Hahotoé, Togo	middle Eocene; Lutetian
Indet.	Eutheria	Kpogamé-Hahotoé, Togo	middle Eocene; Lutetian
Dagbatitherium tassyi Hautier et al., 2021	Elephantiformes; Proboscidea	Kpogamé-Hahotoé, Dagbati Quarry, Togo; KH Phosphate Complex, phospharenite member	middle Eocene; Lutetian
Tanzanycteris mannardi Gunnell et al., 2003	Tanzanycterididae, Eochiroptera; Chiroptera	Mahenge, Tanzania	middle Eocene; Lutetian
Saloumia gorodiskii Tabuce et al., 2020	Moeritheriidae?, Moeritherioidea; Proboscidea	M'Bodione Dadere, Senegal	middle Eocene; Lutetian
Indet.	Paenungulata? ("Condylarthra")	M'Bodione Dadere, Senegal	middle Eocene; Lutetian
Indet.	Eutheria	M'Bodione Dadere, Senegal	middle Eocene; Lutetian
Indet. 1	Eulipotyphla?	Aznag, Ouarzazate Basin, Morocco; Jebel Tagount Fm	middle Eocene; Lutetian
Indet. 2	Eulipotyphla?	Aznag, Ouarzazate Basin, Morocco; Jebel Tagount Fm	middle Eocene; Lutetian
Indet.	Rodentia	Aznag, Ouarzazate Basin, Morocco; Jebel Tagount Fm	middle Eocene; Lutetian

(continued)

Genus and species Author, year	Suprageneric taxonomy	Locality; stratigraphic level	Epoch; age
Indet.	Omomyidae?, Tarsiiformes?; Primates	Aznag, Ouarzazate Basin, Morocco; Jebel Tagount Fm	middle Eocene; Lutetian
Indet. 1	Microchiroptera; Chiroptera	Aznag, Ouarzazate Basin, Morocco; Jebel Tagount Fm	middle Eocene; Lutetian
Indet. 2	Microchiroptera; Chiroptera	Aznag, Ouarzazate Basin, Morocco; Jebel Tagount Fm	middle Eocene; Lutetian
Indet.	Louisinidae? ("Condylarthra"?)	Aznag, Ouarzazate Basin, Morocco; Jebel Tagount Fm	middle Eocene; Lutetian
?Paschatherium sp.	Louisinidae? ("Condylarthra"?)	Aznag, Ouarzazate Basin, Morocco; Jebel Tagount Fm	middle Eocene; Lutetian
Namalestes gheerbranti Pickford et al., 2008	Todralestidae; Eutheria	Black Crow, Sperrgebiet, Namibia	middle Eocene; Lutetian
Indet.	Adapisoriculidae; Eutheria	Black Crow, Sperrgebiet, Namibia	middle Eocene; Lutetian
Indet.	Erinaceidae?; Eulipotyphla?	Black Crow, Sperrgebiet, Namibia	middle Eocene; Lutetian
Pterodon sp.	Hyainailourinae, Hyainailouridae, Hyainailouroidea; Hyaenodonta	Black Crow, Sperrgebiet, Namibia	middle Eocene; Lutetian
Nanogale fragilis Pickford, 2019	Tenrecoidea (= Afrosoricida)	Black Crow, Sperrgebiet, Namibia	middle Eocene; Lutetian
Diamantochloris inconcessus Pickford, 2015	Chrysochloridae; Tenrecoidea (= Afrosoricida)	Black Crow, Sperrgebiet, Namibia	middle Eocene; Lutetian
Indet.	Tenrecoidea (= Afrosoricida)	Black Crow, Sperrgebiet, Namibia	middle Eocene; Lutetian
Damarachloris primaevus Pickford, 2019	Chrysochloridae; Tenrecoidea (= Afrosoricida)	Black Crow, Sperrgebiet, Namibia	middle Eocene; Lutetian
Indet.	Macroscelididae; Macroscelidea	Black Crow, Sperrgebiet, Namibia	middle Eocene; Lutetian

Taxon	Family; Order	Locality	Age
Namahyrax corvus Pickford et al., 2008	Namahyracidae or Geniohyidae?; Hyracoidea	Black Crow, Sperrgebiet, Namibia	middle Eocene; Lutetian
Namatherium blackcrowense Pickford et al., 2008	Arsinoitheriidae; Embrithopoda	Black Crow, Sperrgebiet, Namibia	middle Eocene; Lutetian
Namaparamys inexpectatus Mein and Pickford, 2018	Paramyidae, Reithroparamyinae; Rodentia	Black Crow, Sperrgebiet, Namibia	middle Eocene; Lutetian
Tsaukhaebmys calcareus Pickford, 2018	Zegdoumyidae, Anomaluroidea; Rodentia	Black Crow, Sperrgebiet, Namibia	middle Eocene; Lutetian
Zegdoumys namibiensis (Pickford et al., 2008) Pickford et al., 2018	Zegdoumyidae, Anomaluroidea; Rodentia	Black Crow, Sperrgebiet, Namibia	middle Eocene; Lutetian
Notnamaia bogenfelsi (Pickford et al., 2008) Pickford and Uhen, 2014	Caenopithecinae, Adapidae, Adapiformes, Strepsirrhini; Primates	Black Crow, Sperrgebiet, Namibia	middle Eocene; Lutetian
Namadapis interdictus Godinot et al., 2018	Caenopithecinae, Adapidae, Adapiformes, Strepsirrhini; Primates	Black Crow, Sperrgebiet, Namibia	middle Eocene; Lutetian
Indet.	Propotinae?; Pteropodidae?, Megachiroptera?; Chiroptera	Black Crow, Sperrgebiet, Namibia	middle Eocene; Lutetian
Indet.	Prorastomidae; Sirenia	Taïba Ndiaye, Senegal; Taïba Fm	middle Eocene; Lutetian
Carolinacetus sp.	Protocetidae, Archaeoceti; Cetacea	Taïba Ndiaye, Senegal; Taïba Fm	middle Eocene; Lutetian
Eocetus schweinfurthi (Fraas, 1904) Gingerich, 1992	Basilosauridae, Archaeoceti; Cetacea	Djebel Mokattam (Le Caire), Egypt; Mokattam Limestone, Guishi Fm	middle Eocene; Bartonian
Indet.	Sirenia	Daban, Somalia ("Dalan" in Savage 1969); Nautilus Beds	middle Eocene; Bartonian
Indet.	Hyracoidea	Djebel el Kébar, Tunisia	middle Eocene; Bartonian

(continued)

Genus and species Author, year	Suprageneric taxonomy	Locality; stratigraphic level	Epoch; age
Indet.	Sirenia	Djebel el Kébar, Tunisia	middle Eocene; Bartonian
"Protophiomys" tunisiensis Marivaux et al., 2014	Phiocricetomyinae, Hystricognathi; Rodentia	Djebel el Kébar, Tunisia	middle Eocene; Bartonian
Amamria tunisiensis Marivaux et al., 2014	Simiiformes or Eosimiiformes, Anthropoidea, Primates	Djebel el Kébar, Tunisia	middle Eocene; Bartonian
Barytherium?	Barytheriidae, Barytherioidea; Proboscidea	Gueran, Sahara, Morocco; Aridal Fm	middle Eocene; Bartonian
Platyosphys aithai Gingerich and Zouhri, 2015	Basilosauridae, Archaeoceti; Cetacea	Gueran, Sahara, Morocco; Aridal Fm	middle Eocene; Bartonian
Chrysocetus fouadassii Gingerich and Zouhri, 2015	Basilosauridae, Archaeoceti; Cetacea	Gueran, Sahara, Morocco; Aridal Fm	middle Eocene; Bartonian
Indet. A	Protocetidae, Archaeoceti; Cetacea	Gueran, Sahara, Morocco; Aridal Fm	middle Eocene; Bartonian
Indet. B	Protocetidae, Archaeoceti; Cetacea	Gueran, Sahara, Morocco; Aridal Fm	middle Eocene; Bartonian
Pappocetus lugardi Andrews, 1920	Protocetidae, Archaeoceti; Cetacea	Gueran, Sahara, Morocco; Aridal Fm	middle Eocene; Bartonian
Eocetus schweinfurthi (Fraas, 1904)	Basilosauridae, Archaeoceti; Cetacea	Gueran, Sahara, Morocco; Aridal Fm	middle Eocene; Bartonian
Dorudon atrox? Andrews, 1906 (*"Zeuglodon"* cf. *osiris* in Elouard, 1966)	Basilosauridae, Archaeoceti; Cetacea	Kaolack, Senegal (loc. Tiavandou); "Calcaire à *Nummulites tchiatcheffi*"	middle or late Eocene; Bartonian or Priabonian
Indet.	Eulipotyphla?	Bir El Ater (Nementcha), Algeria	late Eocene; Priabonian
Masrasector cf. *ligabuei* Crochet et al., 1990	Hyainailourinae, Hyainailouridae, Hyainailouroidea; Hyaenodonta	Bir El Ater (Nementcha), Algeria	late Eocene; Priabonian
Indet.	Hyainailourinae, Hyainailouridae, Hyainailouroidea; Hyaenodonta	Bir El Ater (Nementcha), Algeria	late Eocene; Priabonian

Indet.	Hyaenodonta or Carnivoramorpha	Bir El Ater (Nementcha), Algeria	late Eocene; Priabonian
Nementchatherium senarhense Tabuce et al., 2001	Herodotiinae, Macroscelididae; Macroscelidea	Bir El Ater (Nementcha), Algeria	late Eocene; Priabonian
Bunohyrax matsumotoi Tabuce et al., 2000	Geniohyidae; Hyracoidea	Bir El Ater (Nementcha), Algeria	late Eocene; Priabonian
Moeritherium chehbeurameuri Delmer et al., 2006	Moeritheriidae, Moeritherioidea; Proboscidea	Bir El Ater (Nementcha), Algeria	late Eocene; Priabonian
Nementchamys lavocati Jaeger et al., 1985	Nementchamyidae, Anomaluroidea; Rodentia	Bir El Ater (Nementcha), Algeria	late Eocene; Priabonian
Protophiomys algeriensis Jaeger et al., 1985	Protophiomyinae, Hystricognathi; Rodentia	Bir El Ater (Nementcha), Algeria	late Eocene; Priabonian
cf. *Algeripithecus* sp.	Azibiidae, Strepsirrhini; Primates	Bir El Ater (Nementcha), Algeria	late Eocene; Priabonian
Biretia piveteaui De Bonis et al., 1988	Parapithecidae?, Parapithecoidea, Simiiformes, Anthropoidea; Primates	Bir El Ater (Nementcha), Algeria	late Eocene; Priabonian
Indet.	Oligopithecidae?, Catarrhini?, Simiiformes, Anthropoidea; Primates	Bir El Ater (Nementcha), Algeria	late Eocene; Priabonian
Indet.	Anthracotheriidae?	Bir El Ater (Nementcha), Algeria	late Eocene; Priabonian
Arsinoitherium? sp.	Arsinoitheriidae; Embrithopoda	Bir Om Ali, Djebel Chambi, Tunisia	late Eocene; Priabonian
Numidotherium? sp.	Numidotheriidae?; Proboscidea	Dakhla Basin, Sahara, Morocco; Samlat Fm, Itgui–Gerran members, B1 Unit	late Eocene; Priabonian
Dakhlasiren marocensis Zouhri et al., 2022	Protosirenidae; Sirenia	Dakhla Basin, Sahara, Morocco; Samlat Fm, Unit 2 or 3	late Eocene; Priabonian
cf. *Eotheroides* sp.	Dugongidae; Sirenia	Dakhla Basin, Sahara, Morocco; Samlat Fm, Unit 2 or 3	late Eocene; Priabonian

(continued)

Genus and species Author, year	Suprageneric taxonomy	Locality; stratigraphic level	Epoch; age
Dorudon atrox Andrews, 1906	Basilosauridae, Archaeoceti; Cetacea	Dakhla Basin, Sahara, Morocco; Samlat Fm, Itgui–Gerran members, B1 Unit	late Eocene; Priabonian
Basilosaurus isis (Beadnell in Andrews, 1904) Gingerich, 1992	Basilosauridae, Archaeoceti; Cetacea	Dakhla Basin, Sahara, Morocco; Samlat Fm, Itgui–Gerran members, B1 Unit	late Eocene; Priabonian
Basilosaurus sp.	Basilosauridae, Archaeoceti; Cetacea	Dakhla Basin, Sahara, Morocco; Samlat Fm, Itgui–Gerran members, B1 Unit	late Eocene; Priabonian
cf. *Dorudon* sp.	Basilosauridae, Archaeoceti; Cetacea	Dakhla Basin, Sahara, Morocco; Samlat Fm, Itgui–Gerran members, B1 Unit	late Eocene; Priabonian
cf. *Saghacetus* sp.	Basilosauridae, Archaeoceti; Cetacea	Dakhla Basin, Sahara, Morocco; Samlat Fm, Itgui–Gerran members, B1 Unit	late Eocene; Priabonian
cf. *Stromerius* sp.	Basilosauridae, Archaeoceti; Cetacea	Dakhla Basin, Sahara, Morocco; Samlat Fm, Itgui–Gerran members, B1 Unit	late Eocene; Priabonian
Indet.	Proviverrinae?; Hyaenodonta	Dur at Talah, Libya; Idam Unit (="Bioturbated Unit")	late Eocene; Priabonian
Indet.	Hyracoidea	Dur at Talah, Libya; Idam Unit (="Bioturbated Unit")	late Eocene; Priabonian
Barytherium grave Andrews, 1901	Barytheriidae, Barytherioidea; Proboscidea	Dur at Talah, Libya; Idam Unit (="Bioturbated Unit")	late Eocene; Priabonian
Phiomia serridens Andrews and Beadnell, 1902	Palaeomastodontidae, Elephantiformes; Proboscidea	Dur at Talah, Libya; Idam Unit (="Bioturbated Unit")	late Eocene; Priabonian
Moeritherium trigodon Andrews, 1904	Moeritheriidae, Moeritherioidea; Proboscidea	Dur at Talah, Libya; Idam Unit (="Bioturbated Unit")	late Eocene; Priabonian
Indet.	Sirenia	Dur at Talah, Libya; Idam Unit (="Bioturbated Unit")	late Eocene; Priabonian

Taxon	Higher taxonomy	Locality	Age
Indet.	Rodentia	Dur at Talah, Libya; Idam Unit (="Bioturbated Unit")	late Eocene; Priabonian
Arsinoitherium sp.	Arsinoitheriidae; Embrithopoda	Dur at Talah, Libya; Idam Unit (="Bioturbated Unit")	late Eocene; Priabonian
Arcanotherium savagei (Court, 1995) Delmer, 2009	Barytheriidae?, Barytherioidea?; Proboscidea	Dur at Talah, Libya; Evaporite Unit	late Eocene; Priabonian
Phiomia sp.	Palaeomastodontidae, Elephantiformes; Proboscidea,	Dur at Talah, Libya; Evaporite Unit	late Eocene; Priabonian
Moeritherium cf. *lyonsi* Andrews, 1901	Moeritheriidae, Moeritherioidea; Proboscidea	Dur at Talah, Libya; Evaporite Unit	late Eocene; Priabonian
Indet.	Sirenia	Dur at Talah, Libya; Evaporite Unit	late Eocene; Priabonian
Apterodon langebadreae Grohé et al., 2012	Apterodontinae, Hyainailouridae, Hyainailouroidea; Hyaenodonta	Dur at Talah, Libya; several sites in Idam Unit; Idam Unit (="Bioturbated Unit")	late Eocene; Priabonian
Biretia piveteaui De Bonis et al., 1988	Parapithecidae?; Parapithecoidea, Anthropoidea; Primates	Dur at Talah, Libya; DT1 and DT2 loc; Idam Unit (="Bioturbated Unit")	late Eocene; Priabonian
Eotmantsoius perseverans Tabuce et al., 2012	Herodotiinae, Macroscelididae; Macroscelidea	Dur at Talah, Libya; DT1 loc; Idam Unit (="Bioturbated Unit")	late Eocene; Priabonian
Nementchatherium rathbuni Tabuce et al., 2012	Herodotiinae, Macroscelididae; Macroscelidea	Dur at Talah, Libya; DT1 loc; Idam Unit (="Bioturbated Unit")	late Eocene; Priabonian
Arcanotherium savagei (Court, 1995) Delmer, 2009	Barytheriidae?, Barytherioidea?; Proboscidea	Dur at Talah, Libya; DT1 loc; Idam Unit (="Bioturbated Unit")	late Eocene; Priabonian
"*Phiomys*" *hammudai* Jaeger et al., 2010	Phiomyidae, Hystricognathi; Rodentia	Dur at Talah, Libya; DT1 loc; Idam Unit (="Bioturbated Unit")	late Eocene; Priabonian
Talahphiomys libycus Jaeger et al., 2010	Phiocricetomyinae, Hystricognathi; Rodentia	Dur at Talah, Libya; DT1 loc; Idam Unit (="Bioturbated Unit")	late Eocene; Priabonian

(continued)

Genus and species Author, year	Suprageneric taxonomy	Locality; stratigraphic level	Epoch; age
Afrotarsius libycus Jaeger et al., 2010	Afrotarsiidae, Eosimiiformes? or Tarsiiformes?; Primates	Dur at Talah, Libya; DT1 loc; Idam Unit (="Bioturbated Unit")	late Eocene; Priabonian
Talahpithecus parvus Jaeger et al., 2010	Oligopithecidae; Catarrhini, Anthropoidea; Primates	Dur at Talah, Libya; DT1 loc; Idam Unit (="Bioturbated Unit")	late Eocene; Priabonian
Karanisia arenula Jaeger et al., 2010	Lorsiformes, Strepsirrhini; Primates	Dur at Talah, Libya; DT1 loc; Idam Unit (="Bioturbated Unit")	late Eocene; Priabonian
"Protophiomys" durattalahensis Jaeger et al., 2010	Protophiomyinae, Hystricognathi; Rodentia	Dur at Talah, Libya; DT2 loc; Idam Unit (="Bioturbated Unit")	late Eocene; Priabonian
Talahphiomys lavocati (Wood, 1968) Coster et al., 2012	Phiocricetomyinae, Hystricognathi; Rodentia	Dur at Talah, Libya; DT2 loc; Idam Unit (="Bioturbated Unit")	late Eocene; Priabonian
Kabirmys prius Coster et al., 2015	Nementchamyidae, Anomaluroidea; Rodentia	Dur at Talah, Libya; DT3 loc; Idam Unit (="Bioturbated Unit")	late Eocene; Priabonian
Apterodon Indet.	Apterodontinae, Hyainailouridae, Hyainailouroidea; Hyaenodonta	Dur at Talah, Libya; Idam Unit (="Bioturbated Unit")	late Eocene; Priabonian
Nementchamys lavocati Jaeger et al., 1985	Nementchamyidae, Anomaluroidea; Rodentia	Dur at Talah, Libya; Idam Unit (="Bioturbated Unit")	late Eocene; Priabonian
Basilosaurus isis (Beadnell in Andrews, 1904) Gingerich, 1992	Basilosauridae, Archaeoceti; Cetacea	Fayum W (Birket Qarun cliffs), Egypt; Wadi Al-Hitan, Birket Qarun Fm	late Eocene; Priabonian
Dorudon atrox Andrews, 1906	Basilosauridae, Archaeoceti; Cetacea	Fayum W (Garand Gehannam), Egypt; Wadi Al-Hitan (Gingerich and Uhen, 1996)	late Eocene; Priabonian
Protosiren sp.	Protosirenidae; Sirenia	Fayum W, Egypt; Wadi Al-Hitan	middle or late Eocene; Bartonian or Priabonian
Pontogeneus? brachyspondylus (Müller, 1849) Gingerich, 1992	Basilosauridae, Archaeoceti; Cetacea	Fayum W, Egypt; Wadi Al-Hitan	late Eocene; Priabonian

Taxon	Family, higher taxa	Locality	Age
Ancalecetus simonsi Gingerich and Uhen, 1996	Basilosauridae, Archaeoceti; Cetacea	Fayum W, Egypt; Wadi Al-Hitan (Gingerich & Uhen 1996)	late Eocene; Priabonian
Aegicetus gehennae Gingerich et al., 2019	Protocetidae, Archaeoceti; Cetacea	Fayum W, Egypt; Wadi Al-Hitan; Gehannam Fm (base)	late Eocene; Priabonian
Protosiren smithae Domning and Gingerich, 1994	Protosirenidae; Sirenia	Fayum W, Egypt; Wadi Al-Hitan, loc. ZV-54; Birket Qarun Fm	late Eocene; Priabonian,
Saghacetus osiris (Dames, 1894) Gingerich, 1992	Basilosauridae, Archaeoceti; Cetacea	Fayum, Egypt; Qasr El Sagha Fm	late Eocene; Priabonian
Moeritherium lyonsi Andrews, 1901	Moeritheriidae, Moeritherioidea; Proboscidea	Fayum, Egypt; Qasr El Sagha Fm, Dir Abu Lifa Member	late Eocene; Priabonian
Apterodon sp.	Apterodontinae, Hyainailouridae, Hyainailouroidea; Hyaenodonta	Fayum, Egypt; Qasr El Sagha Fm, Dir Abu Lifa Member	late Eocene; Priabonian
Pterodon sp.	Hyainailourinae, Hyainailouridae, Hyainailouroidea; Hyaenodonta	Fayum, Egypt; Qasr El Sagha Fm, Dir Abu Lifa Member	late Eocene; Priabonian
Apterodon saghensis Simons and Gingerich, 1976	Apterodontinae, Hyainailouridae, Hyainailouroidea; Hyaenodonta	Fayum, Egypt; Qasr El Sagha Fm, Dir Abu Lifa Member	late Eocene; Priabonian
Metapterodon cf. *brachycephalus* Osborn, 1909	Hyainailourinae, Hyainailouridae, Hyainailouroidea; Hyaenodonta	Fayum, Egypt; Qasr El Sagha Fm, Dir Abu Lifa Member	late Eocene; Priabonian
Indet.	Saghatheriidae?; Hyracoidea	Fayum, Egypt; Qasr El Sagha Fm, Dir Abu Lifa Member	late Eocene; Priabonian
Arsinoitherium sp.	Arsinoitheriidae; Embrithopoda	Fayum, Egypt; Qasr El Sagha Fm, Dir Abu Lifa Member	late Eocene; Priabonian
Barytherium grave Andrews, 1901	Barytheriidae, Barytherioidea; Proboscidea	Fayum, Egypt; Qasr El Sagha Fm, Dir Abu Lifa Member	late Eocene; Priabonian
Barytherium n. sp.	Barytheriidae, Barytherioidea; Proboscidea	Fayum, Egypt; Qasr El Sagha Fm, Dir Abu Lifa Member	late Eocene; Priabonian

(continued)

Genus and species Author, year	Suprageneric taxonomy	Locality; stratigraphic level	Epoch; age
Eosiren libyca Andrew, 1902	Halitheriinae, Dugongidae; Sirenia	Fayum, Egypt; Qasr El Sagha Fm, Dir Abu Lifa Member	late Eocene; Priabonian
Eosiren stromeri (Sickenberg, 1934) Kordos, 1977	Halitheriinae, Dugongidae; Sirenia	Fayum, Egypt; Qasr El Sagha Fm, Dir Abu Lifa Member	late Eocene; Priabonian
Bothriogenys sp.	Anthracotheriidae, Hippopotamoidea; Artiodactyla	Fayum, Egypt; Qasr El Sagha Fm, Dir Abu Lifa Member	late Eocene; Priabonian
Dorudon stromeri (Kellog, 1928) Gingerich, 1992	Basilosauridae, Archaeoceti; Cetacea	Fayum, Egypt; Qasr El Sagha Fm, Dir Abu Lifa Member	late Eocene; Priabonian
Ghamidtherium dimaiensis Sanchez-Villagra et al., 2007	Didelphimorphia? Marsupialia?	BQ2, Fayum, Egypt; Birket Qarun Fm, Umm Rigl Member	late Eocene; Priabonian
Indet. (new)	Kelbidae?; Ptolemaiida	BQ2, Fayum, Egypt; Birket Qarun Fm, Umm Rigl Member	late Eocene; Priabonian
Masrasector? sp.	Teratodontinae, Hyainailouridae, Hyainailouroidea; Hyaenodonta	BQ2, Fayum, Egypt; Birket Qarun Fm, Umm Rigl Member	late Eocene; Priabonian
Dilambdogale gheerbranti Seiffert, 2010	Tenrecoidea? (= Afrosoricida?)	BQ2, Fayum, Egypt; Birket Qarun Fm, Umm Rigl Member	late Eocene; Priabonian
Indet. (spp.)	Herodotiinae, Macroscelididae; Macroscelidea	BQ2, Fayum, Egypt; Birket Qarun Fm, Umm Rigl Member	late Eocene; Priabonian
Bunohyrax matsumotoi Tabuce et al., 2000	Geniohyidae; Hyracoidea	BQ2, Fayum, Egypt; Birket Qarun Fm, Umm Rigl Member	late Eocene; Priabonian
Dimaitherium patnaiki Barrow et al., 2010	Hyracoidea	BQ2, Fayum, Egypt; Birket Qarun Fm, Umm Rigl Member	late Eocene; Priabonian
Nementchamys lavocati Jaeger et al., 1985	Nementchamyidae, Anomaluroidea; Rodentia	BQ2, Fayum, Egypt; Birket Qarun Fm, Umm Rigl Member	late Eocene; Priabonian

Shazurus minutus Sallam et al., 2010	Anomaluridae?, Anomaluroidea; Rodentia	BQ2, Fayum, Egypt; Birket Qarun Fm, Umm Rigl Member	late Eocene; Priabonian
Kabirmys qarunensis Sallam et al., 2010	Nementchamyidae, Anomaluroidea; Rodentia	BQ2, Fayum, Egypt; Birket Qarun Fm, Umm Rigl Member	late Eocene; Priabonian
Protophiomys aegyptensis Sallam et al., 2009	Protophiomyinae, Hystricognathi; Rodentia	BQ2, Fayum, Egypt; Birket Qarun Fm, Umm Rigl Member	late Eocene; Priabonian
Waslamys attiai Sallam et al., 2009	Phiocricetomyinae, Hystricognathi; Rodentia	BQ2, Fayum, Egypt; Birket Qarun Fm, Umm Rigl Member	late Eocene; Priabonian
Nosmips aenigmaticus Seiffert et al., 2010	Primates	BQ2, Fayum, Egypt; Birket Qarun Fm, Umm Rigl Member	late Eocene; Priabonian
Biretia fayumensis Seiffert et al., 2005	Parapithecidae?, Parapithecoidea, Anthropoidea; Primates	BQ2, Fayum, Egypt; Birket Qarun Fm, Umm Rigl Member	late Eocene; Priabonian
Biretia megalopsis Seiffert et al., 2005	Parapithecoidea?, Parapithecoidea, Anthropoidea; Primates	BQ2, Fayum, Egypt; Birket Qarun Fm, Umm Rigl Member	late Eocene; Priabonian
Indet.	Djebelemuridae?, Strepsirrhini; Primates	BQ2, Fayum, Egypt; Birket Qarun Fm, Umm Rigl Member	late Eocene; Priabonian
Afradapis longicristatus Seiffert et al., 2009	Adapidae, Adapiformes, Strepsirrhini; Primates	BQ2, Fayum, Egypt; Birket Qarun Fm, Umm Rigl Member	late Eocene; Priabonian
Masradapis tahai Seiffert et al., 2018	Caenopithecinae, Adapidae, Adapiformes, Strepsirrhini; Primates	BQ2, Fayum, Egypt; Birket Qarun Fm, Umm Rigl Member	late Eocene; Priabonian
Saharagalago misrensis Seiffert et al., 2003	Galagidae, Lorisiformes Strepsirrhini; Primates	BQ2, Fayum, Egypt; Birket Qarun Fm, Umm Rigl Member	late Eocene; Priabonian
Karanisia clarki Seiffert et al., 2003	Lorisidae?, Lorisiformes?, Strepsirrhini, Primates	BQ2, Fayum, Egypt; Birket Qarun Fm, Umm Rigl Member	late Eocene; Priabonian

(continued)

Genus and species Author, year	Suprageneric taxonomy	Locality; stratigraphic level	Epoch; age
Phasmatonycteris butleri Gunnell et al., 2014	Myzopodidae; Microchiroptera; Chiroptera	BQ2, Fayum, Egypt; Birket Qarun Fm, Umm Rigl Member	late Eocene; Priabonian
Witwatia eremicus Gunnell et al., 2008	Philisidae, Microchiroptera; Chiroptera	BQ2, Fayum, Egypt; Birket Qarun Fm, Umm Rigl Member	late Eocene; Priabonian
Witwatia schlosseri Gunnell et al., 2008	Philisidae; Microchiroptera; Chiroptera	BQ2, Fayum, Egypt; Birket Qarun Fm, Umm Rigl Member	late Eocene; Priabonian
Aegyptonycteris knightae Simmons et al., 2016	Aegyptonycteridae, Microchiroptera; Chiroptera	BQ2, Fayum, Egypt; Birket Qarun Fm, Umm Rigl Member	late Eocene; Priabonian
Qarunycteris moerisae Gunnell et al., 2008	Rhinopomatidae, Microchiroptera; Chiroptera	BQ2, Fayum, Egypt; Birket Qarun Fm, Umm Rigl Member	late Eocene; Priabonian
Moeritherium sp.	Moeritheriidae, Moeritherioidea; Proboscidea	Birket Qarun loc. 2–BQ2, loc L-67, Fayum, Egypt; Birket Qarun Fm, Umm Rigl Member	late Eocene; Priabonian
Moeritherium sp.	Moeritheriidae, Moeritherioidea; Proboscidea	In Tafidet, Mali	late Eocene; Priabonian
Moeritherium sp.	Moeritheriidae, Moeritherioidea; Proboscidea	Khenchela (= Bir El Ater?), Algeria	late Eocene; Priabonian
Indet.	Sirenia?	Bedeil, Somalia	early Oligocene; Rupelian
Nonanomalurus parvus Marivaux et al., 2017	Nonanomaluridae, Anomaluroidea; Rodentia	Dakhla Basin (El Argoub and Porto Rico loc.), Sahara, Morocco; Upper Samlat Fm, U4 level C2	early Oligocene; Rupelian
Paranomalurus riodeoroensis Marivaux et al., 2017	Anomaluridae, Anomaluroidea; Rodentia	Dakhla Basin (El Argoub and Porto Rico loc.), Sahara, Morocco; Upper Samlat Fm, U4 level C2	early Oligocene; Rupelian

Birkamys aff. *korai* Sallam and Seiffert, 2016	Phiocricetomyinae, Hystricognathi; Rodentia	Dakhla Basin (El Argoub and Porto Rico loc.), Sahara, Morocco; Upper Samlat Fm, U4 level C2	early Oligocene; Rupelian
Argouburus minutus Marivaux et al., 2017	Anomaluridae, Anomaluroidea; Rodentia	Dakhla Basin (El Argoub loc.), Sahara, Morocco; Upper Samlat Fm, U4 level C2	early Oligocene; Rupelian
Oromys zenkerellinopsis Marivaux et al., 2017	Zenkerellidae; Anomaluroidea; Rodentia	Dakhla Basin (El Argoub loc.), Sahara, Morocco; Upper Samlat Fm, U4 level C2	early Oligocene; Rupelian
Mubhammys atlanticus Marivaux et al., 2017	Phiocricetomyinae, Hystricognathi; Rodentia	Dakhla Basin (El Argoub loc.), Sahara, Morocco; Upper Samlat Fm, U4 level C2	early Oligocene; Rupelian
Gaudeamus cf. *aslius* Sallam et al., 2011	Gaudeamuridae, Hystricognathi; Rodentia	Dakhla Basin (El Argoub loc.), Sahara, Morocco; Upper Samlat Fm, U4 level C2	early Oligocene; Rupelian
Phenacophiomys occidentalis Marivaux et al., 2017	Phiocricetomyinae, Hystricognathi; Rodentia	Dakhla Basin (El Argoub loc.), Sahara, Morocco; Upper Samlat Fm, U4 level C2	early Oligocene; Rupelian
Dakhlamys ultimus Marivaux et al., 2017	Zegdoumyidae?, Anomaluroidea; Rodentia	Dakhla Basin (Porto Rico loc.), Sahara, Morocco; Upper Samlat Fm, U4 level C2	early Oligocene; Rupelian
Gaudeamus cf. *hylaeus* Sallam et al., 2011	Gaudeamuridae; Rodentia Hystricognathi	Dakhla Basin (Porto Rico loc.), Sahara, Morocco; Upper Samlat Fm, U4 level C2	early Oligocene; Rupelian
Neophiomys minutus Marivaux et al., 2017	Phiocricetomyinae, Hystricognathi; Rodentia	Dakhla Basin (Porto Rico loc.), Sahara, Morocco; Upper Samlat Fm, U4 level C2	early Oligocene; Rupelian
Phiocricetomys sp.	Phiocricetomyinae, Hystricognathi; Rodentia	Dakhla Basin (Porto Rico loc.), Sahara, Morocco; Upper Samlat Fm, U4 level C2	early Oligocene; Rupelian

(continued)

Genus and species Author, year	Suprageneric taxonomy	Locality; stratigraphic level	Epoch; age
Indet.	Sirenia	Djebel ech Cherichira, Tunisia (= Jebel Bon Gobrine?); Grès de Fortuna	early Oligocene; Rupelian
Qarunavus meyeri Simons and Gingerich, 1974	Ptolemaiidae; Ptolemaiida	Fayum, Egypt; Jebel Qatrani Fm	early Oligocene; Rupelian
Indet.	Eulipotyphla?	Fayum, Egypt; Jebel Qatrani Fm	early Oligocene; Rupelian
Metasinopa aethiopica (Andrews, 1906)	Teratodontinae, Hyainailouridae, Hyainailouroidea; Hyaenodonta	Fayum, Egypt; Jebel Qatrani Fm	early Oligocene; Rupelian
Pterodon africanus Andrews, 1903	Apterodontinae, Hyainailouridae, Hyainailouroidea; Hyaenodonta	Fayum, Egypt; Jebel Qatrani Fm	early Oligocene; Rupelian
Apterodon altidens Schlosser, 1910	Apterodontinae, Hyainailouridae, Hyainailouroidea; Hyaenodonta	Fayum, Egypt; Jebel Qatrani Fm	early Oligocene; Rupelian
Metapterodon brachycephalus Osborn, 1909	Hyainailourinae, Hyainailouridae, Hyainailouroidea; Hyaenodonta	Fayum, Egypt; Jebel Qatrani Fm	early Oligocene; Rupelian
Apterodon macrognathus (Andrews, 1904) Andrews, 1906	Apterodontinae, Hyainailouridae, Hyainailouroidea; Hyaenodonta	Fayum, Egypt; Jebel Qatrani Fm	early Oligocene; Rupelian
Quasiapterodon minutus (Schlosser, 1910) Lavrov, 1999	Apterodontinae, Hyainailouridae, Hyainailouroidea; Hyaenodonta	Fayum, Egypt; Jebel Qatrani Fm	early Oligocene; Rupelian
Metoldobotes stromeri Schlosser, 1910	Metoldobotinae, Macroscelididae; Macroscelidea	Fayum, Egypt; Jebel Qatrani Fm	early Oligocene; Rupelian
Saghatherium "sobrina" Matsumoto, 1926	Saghatheriidae; Hyracoidea	Fayum, Egypt; Jebel Qatrani Fm	early Oligocene; Rupelian
Geniohyus diphycus Matsumoto, 1926	Geniohyidae; Hyracoidea	Fayum, Egypt; Jebel Qatrani Fm	early Oligocene; Rupelian
Megalohyrax eocaenus Andrews, 1903	Geniohyidae; Hyracoidea	Fayum, Egypt; Jebel Qatrani Fm	early Oligocene; Rupelian
Bunohyrax fajumensis Andrews, 1904	Geniohyidae; Hyracoidea	Fayum, Egypt; Jebel Qatrani Fm	early Oligocene; Rupelian

Bunohyrax major (Andrews, 1904) Schlosser, 1910	Geniohyidae; Hyracoidea	Fayum, Egypt; Jebel Qatrani Fm	early Oligocene; Rupelian
Geniohyus mirus Andrews, 1904	Geniohyidae; Hyracoidea	Fayum, Egypt; Jebel Qatrani Fm	early Oligocene; Rupelian
Titanohyrax ultimus Matsumoto, 1922	Titanohyracidae; Hyracoidea	Fayum, Egypt; Jebel Qatrani Fm	early Oligocene; Rupelian
Arsinoitherium andrewsi Lankester, 1903	Arsinoitheriidae; Embrithopoda	Fayum, Egypt; Jebel Qatrani Fm	early Oligocene; Rupelian
Arsinoitherium zitelli Beadnell, 1902	Arsinoitheriidae; Embrithopoda	Fayum, Egypt; Jebel Qatrani Fm	early Oligocene; Rupelian
Palaeomastodon beadnelli Andrews, 1901	Palaeomastodontidae, Elephantiformes; Proboscidea	Fayum, Egypt; Jebel Qatrani Fm	early Oligocene; Rupelian
Phiomia serridens Andrews and Beadnell, 1902	Palaeomastodontidae, Elephantiformes; Proboscidea	Fayum, Egypt; Jebel Qatrani Fm	early Oligocene; Rupelian
Moeritherium trigodon Andrews, 1904	Moeritheriidae, Moeritherioidea; Proboscidea	Fayum, Egypt; Jebel Qatrani Fm	early Oligocene; Rupelian
Indet.	Anomaluridae?, Anomaluroidea; Rodentia	Fayum, Egypt; Jebel Qatrani Fm	early Oligocene; Rupelian
Gaudeamus aegyptius Wood, 1968	Gaudeamuridae, Hystricognathi; Rodentia	Fayum, Egypt; Jebel Qatrani Fm	early Oligocene; Rupelian
Phiomys lavocati Wood, 1968	Phiomyidae, Hystricognathi; Rodentia	Fayum, Egypt; Jebel Qatrani Fm	early Oligocene; Rupelian
Vampyravus orientalis Schlosser, 1910	Microchiroptera; Chiroptera	Fayum, Egypt; Jebel Qatrani Fm	early Oligocene; Rupelian
Nabotherium aegyptiacum (Andrews 1906) Sileem et al., 2016	Anthracotheriidae, Hippopotamoidea; Artiodactyla	Fayum, Egypt; Jebel Qatrani Fm	early Oligocene; Rupelian
Bothriogenys andrewsi Schmidt, 1913	Anthracotheriidae, Hippopotamoidea; Artiodactyla	Fayum, Egypt; Jebel Qatrani Fm	early Oligocene; Rupelian

(continued)

Genus and species Author, year	Suprageneric taxonomy	Locality; stratigraphic level	Epoch; age
Bothriogenys gorringei Andrews and Beadnell, 1902	Anthracotheriidae, Hippopotamoidea; Artiodactyla	Fayum, Egypt; Jebel Qatrani Fm	early Oligocene; Rupelian
Qatraniodon parvus (Andrews, 1906) Ducrocq, 1997	Anthracotheriidae, Hippopotamoidea; Artiodactyla	Fayum, Egypt; Jebel Qatrani Fm	early Oligocene; Rupelian
Bothriogenys rugulosus Schmidt, 1913	Anthracotheriidae, Hippopotamoidea; Artiodactyla	Fayum, Egypt; Jebel Qatrani Fm	early Oligocene; Rupelian
Philisis sp.	Philisidae, Vespertilionoidea, Microchiroptera; Chiroptera	Fayum, Egypt; unknown loc., Jebel Qatrani Fm	early Oligocene; Rupelian
Bothriogenys fraasi Schmidt, 1913	Anthracotheriidae, Hippopotamoidea; Artiodactyla	Fayum, Egypt; Quarries I, M, O, Jebel Qatrani Fm	early Oligocene; Rupelian
Pterodon africanus Andrews, 1903	Hyainailourinae, Hyainailouridae, Hyainailouroidea; Hyaenodonta	L41, Fayum, Egypt; Jebel Qatrani Fm	earliest Oligocene; Rupelian (base)
Apterodon altidens Schlosser, 1910	Apterodontinae, Hyainailouridae, Hyainailouroidea; Hyaenodonta	L41, Fayum, Egypt; Jebel Qatrani Fm	earliest Oligocene; Rupelian (base)
Brychotherium ephalmos Borth et al., 2016	Teratodontinae, Hyainailouridae, Hyainailouroidea; Hyaenodonta	L41, Fayum, Egypt; Jebel Qatrani Fm	earliest Oligocene; Rupelian (base)
Akhnatenavus leptognathus (Osborn, 1909) Holroyd, 1999	Hyainailourinae, Hyainailouridae, Hyainailouroidea; Hyaenodonta	L41, Fayum, Egypt; Jebel Qatrani Fm	earliest Oligocene; Rupelian (base)
Akhnatenavus nefertiticyon Borth et al., 2016	Hyainailourinae, Hyainailouridae, Hyainailouroidea; Hyaenodonta	L41, Fayum, Egypt; Jebel Qatrani Fm	earliest Oligocene; Rupelian (base)
Masrasector ligabuei Crochet et al., 1990	Teratodontinae, Hyainailouridae, Hyainailouroidea; Hyaenodonta	L41, Fayum, Egypt; Jebel Qatrani Fm	earliest Oligocene; Rupelian (base)
Apterodon macrognathus Andrews, 1904	Apterodontinae, Hyainailouridae, Hyainailouroidea; Hyaenodonta	L41, Fayum, Egypt; Jebel Qatrani Fm	earliest Oligocene; Rupelian (base)

Taxon	Higher taxonomy	Locality	Age
Masrasector nananubis Borth and Seiffert, 2017	Teratodontinae, Hyainailouridae, Hyainailouroidea; Hyaenodonta	L41, Fayum, Egypt; Jebel Qatrani Fm	earliest Oligocene; Rupelian (base)
Pterodon phiomensis Osborn, 1909	Hyainailourinae, Hyainailouridae, Hyainailouroidea; Hyaenodonta	L41, Fayum, Egypt; Jebel Qatrani Fm	earliest Oligocene; Rupelian (base)
Widanelfarasia bowni Seiffert and Simons, 2000	Tenrecoidea? (= Afrosoricida?)	L41, Fayum, Egypt; Jebel Qatrani Fm	earliest Oligocene; Rupelian (base)
Widanelfarasia rasmusseni Seiffert et al., 2007	Tenrecoidea? (= Afrosoricida?)	L41, Fayum, Egypt; Jebel Qatrani Fm	earliest Oligocene; Rupelian (base)
Herodotius pattersoni Simons et al., 1991	Herodotiinae, Macroscelididae; Macroscelidea	L41, Fayum, Egypt; Jebel Qatrani Fm	earliest Oligocene; Rupelian (base)
Saghatherium bowni Rasmussen and Simons, 1991	Saghatheriidae; Hyracoidea	L41, Fayum, Egypt; Jebel Qatrani Fm	earliest Oligocene; Rupelian (base)
Thyrohyrax litholagus Rasmussen and Simons, 1991	Saghatheriidae; Hyracoidea	L41, Fayum, Egypt; Jebel Qatrani Fm	earliest Oligocene; Rupelian (base)
Thyrohyrax meyeri Rasmussen and Simons, 1991	Saghatheriidae; Hyracoidea	L41, Fayum, Egypt; Jebel Qatrani Fm	earliest Oligocene; Rupelian (base)
Antilohyrax pectidens Rasmussen and Simons, 2000	Titanohyracidae; Hyracoidea	L41, Fayum, Egypt; Jebel Qatrani Fm	earliest Oligocene; Rupelian (base)
Gaudeamus aff. *hylaeus*	Gaudeamuridae, Hystricognathi; Rodentia	L41, Fayum, Egypt; Jebel Qatrani Fm	earliest Oligocene; Rupelian (base)
Gaudeamus aslius Sallam et al., 2011	Gaudeamuridae, Hystricognathi; Rodentia	L41, Fayum, Egypt; Jebel Qatrani Fm	earliest Oligocene; Rupelian (base)
Gaudeamus hylaeus Sallam et al., 2011	Gaudeamuridae, Hystricognathi; Rodentia	L41, Fayum, Egypt; Jebel Qatrani Fm	earliest Oligocene; Rupelian (base)
Birkamys korai Sallam and Seiffert, 2016	Phiocricetomyinae, Hystricognathi; Rodentia	L41, Fayum, Egypt; Jebel Qatrani Fm	earliest Oligocene; Rupelian (base)

(continued)

Genus and species Author, year	Suprageneric taxonomy	Locality; stratigraphic level	Epoch; age
cf. *Birkamys* sp.	Phiocricetomyinae, Hystricognathi; Rodentia	L41, Fayum, Egypt; Jebel Qatrani Fm	earliest Oligocene; Rupelian (base)
Mubhammys vadumensis Sallamet Seiffert, 2016	Phiocricetomyinae, Hystricognathi; Rodentia	L41, Fayum, Egypt; Jebel Qatrani Fm	earliest Oligocene; Rupelian (base)
Acritophiomys bowni Sallam et al., 2012	Metaphiomyinae, Hystricognathi; Rodentia	L41, Fayum, Egypt; Jebel Qatrani Fm	earliest Oligocene; Rupelian (base)
Catopithecus browni Simons, 1989	Oligopithecidae; Catarrhini, Simiiformes, Anthropoidea; Primates	L41, Fayum, Egypt; Jebel Qatrani Fm	earliest Oligocene; Rupelian (base)
Proteopithecus sylviae Simons, 1989	Proteopithecidae; Catarrhini, Simiiformes, Anthropoidea; Primates	L41, Fayum, Egypt; Jebel Qatrani Fm	earliest Oligocene; Rupelian (base)
Abuqatrania basiodontos Simons et al., 2001	Parapithecidae; Parapithecoidea, Simiiformes, Anthropoidea; Primates	L41, Fayum, Egypt; Jebel Qatrani Fm	earliest Oligocene; Rupelian (base)
Serapia eocaena Simons, 1992	Proteopithecidae; Parapithecoidea, Simiiformes, Anthropoidea; Primates	L41, Fayum, Egypt; Jebel Qatrani Fm	earliest Oligocene; Rupelian (base)
Arsinoea kallimos Simons, 1992	Parapithecidae?; Parapithecoidea, Simiiformes, Anthropoidea; Primates	L41, Fayum, Egypt; Jebel Qatrani Fm	earliest Oligocene; Rupelian (base)
Aframonius dieides Simons et al., 1995	Adapidae, Cercamoniinae, Strepsirrhini; Primates	L41, Fayum, Egypt; Jebel Qatrani Fm	earliest Oligocene; Rupelian (base)
Anchomomys milleri Simons, 1997	Djebelemuridae, Strepsirrhini; Primates	L41, Fayum, Egypt; Jebel Qatrani Fm	earliest Oligocene; Rupelian (base)

Plesiopithecus teras Simons, 1992	Plesiopithecidae, Chiromyiformes, Strepsirrhini; Primates	L41, Fayum, Egypt; Jebel Qatrani Fm	earliest Oligocene; Rupelian (base)
Wadilemur elegans Simons, 1997	Galagidae, Lorisiformes, Strepsirrhini; Primates	L41, Fayum, Egypt; Jebel Qatrani Fm	earliest Oligocene; Rupelian (base)
Khonsunycteris aegypticus Gunnell et al., 2008	Vespertilionidae, Vespertilionoidae, Microchiroptera; Chiroptera	L41, Fayum, Egypt; Jebel Qatrani Fm	earliest Oligocene; Rupelian (base)
Saharaderma pseudovampyrus Gunnell et al., 2008	Megadermatidae, Emballonuroidea, Microchiroptera; Chiroptera	L41, Fayum, Egypt; Jebel Qatrani Fm	earliest Oligocene; Rupelian (base)
Dhofarella sigei Gunnell et al., 2008	Emballonuridae, Emballonuroidea, Microchiroptera; Chiroptera	L41, Fayum, Egypt; Jebel Qatrani Fm	earliest Oligocene; Rupelian (base)
Parapithecus fraasi Schlosser, 1911	Parapithecinae, Parapithecidae, Parapithecoidea, Simiiformes, Anthropoidea; Primates	Fayum, Egypt; unknown loc. (Quarry M?), Jebel Qatrani Fm	early Oligocene; Rupelian
Titanohyrax angustidens Rasmussen and Simons, 1988	Titanohyracidae; Hyracoidea	Fayum, Egypt; unknown loc. and Quarries R, V, I, Jebel Qatrani Fm	early Oligocene; Rupelian
Propliopithecus haeckeli Schlosser, 1911	Propliopithecidae; Catarrhini, Simiiformes, Anthropoidea; Primates	Fayum, Egypt; unknown type-loc. (referred material from Quarry G), Jebel Qatrani Fm	early Oligocene; Rupelian
Moeripithecus markgrafi Schlosser, 1910	Propliopithecidae; Catarrhini, Simiiformes, Anthropoidea; Primates	Fayum, Egypt; unknown type-loc., Quarry V, Jebel Qatrani Fm	early Oligocene; Rupelian
Saghatherium antiquum Andrews and Beadnell, 1902	Saghatheriidae; Hyracoidea	Fayum, Egypt; Quarries A, B, E, Jebel Qatrani Fm	early Oligocene; Rupelian
Apidium moustafai Simons, 1962	Parapithecinae, Parapithecidae, Parapithecoidea, Simiiformes, Anthropoidea; Primates	Fayum, Egypt, Quarries G, ?V, Jebel Qatrani Fm	early Oligocene; Rupelian

(continued)

Genus and species Author, year	Suprageneric taxonomy	Locality; stratigraphic level	Epoch; age
Propliopithecus ankeli Simons et al., 1987	Propliopithecidae; Catarrhini, Simiiformes, Anthropoidea; Primates	Fayum, Egypt; Quarries G, V, Jebel Qatrani Fm	early Oligocene; Rupelian
Metaphiomys beadnelli Obsorn, 1908	Metaphiomyinae, Hystricognathi; Rodentia	Fayum, Egypt; Quarries I, J, M, O, P, R, Jebel Qatrani Fm	early Oligocene; Rupelian
Thyrohyrax domorictus Meyer, 1973	Saghatheriidae; Hyracoidea	Fayum, Egypt; Quarries I, M, Jebel Qatrani Fm	early Oligocene; Rupelian
Propliopithecus chirobates Simons, 1965	Propliopithecidae; Catarrhini, Simiiformes, Anthropoidea; Primates	Fayum, Egypt; Quarries I, M, Jebel Qatrani Fm	early Oligocene; Rupelian
Aegyptopithecus zeuxis Simons, 1965	Propliopithecidae; Catarrhini, Simiiformes, Anthropoidea; Primates	Fayum, Egypt, Quarries I, M, Jebel Qatrani Fm	early Oligocene; Rupelian
Pachyhyrax crassidentatus Schlosser, 1910	Titanohyracidae; Hyracoidea	Fayum, Egypt; Quarries I, M, L-46, Jebel Qatrani Fm	early Oligocene; Rupelian
Parapithecus grangeri Simons, 1974	Parapithecinae, Parapithecidae; Parapithecoidea, Simiiformes, Anthropoidea; Primates	Fayum, Egypt; Quarries I, M, R, Jebel Qatrani Fm	early Oligocene; Rupelian
Apidium phiomense Osborn, 1908	Parapithecinae, Parapithecidae, Parapithecoidea, Simiiformes, Anthropoidea; Primates	Fayum, Egypt; Quarries J, I, M, P, Jebel Qatrani Fm	early Oligocene; Rupelian
Afrotarsius chatrathi Simons and Bown, 1985	Afrotarsiidae, Eosimiiformes? or Tarsiiformes?; Primates,	Fayum, Egypt; Quarries M, P, Jebel Qatrani Fm	early Oligocene; Rupelian
Ptolemaia lyonsi Osborn, 1908	Ptolemaiidae; Ptolemaiida	Fayum, Egypt; Quarry A, Jebel Qatrani Fm	early Oligocene; Rupelian

Akhnatenavus leptognathus (Osborn, 1909) Holroyd, 1999	Hyainailourinae, Hyainailouridae, Hyainailouroidea; Hyaenodonta	Fayum, Egypt; Quarry A, Jebel Qatrani Fm	early Oligocene; Rupelian
Pterodon phiomensis Osborn, 1909	Hyainailourinae, Hyainailouridae, Hyainailouroidea; Hyaenodonta	Fayum, Egypt; Quarry A, Jebel Qatrani Fm	early Oligocene; Rupelian
Phiomys andrewsi Osborn, 1908	Phiomyidae, Hystricognathi; Rodentia	Fayum, Egypt; Quarry A, Jebel Qatrani Fm	early Oligocene; Rupelian
Titanohyrax andrewsi Matsumoto, 1922	Titanohyracidae; Hyracoidea	Fayum, Egypt; Quarry A, B, C, Jebel Qatrani Fm	early Oligocene; Rupelian
Thyrohyrax pygmaeus Matsumoto, 1922	Saghatheriidae; Hyracoidea	Fayum, Egypt; Quarry A-1, Jebel Qatrani Fm	early Oligocene; Rupelian
Thyrohyrax sp.	Saghatheriidae; Hyracoidea	Fayum, Egypt; Quarry A-1, Jebel Qatrani Fm	early Oligocene; Rupelian
Jawharia tenrecoides Seiffert et al., 2007	Tenrecoidea? (= Afrosoricida?)	Fayum, Egypt; Quarry E, Jebel Qatrani Fm	early Oligocene; Rupelian
Eochrysochloris tribosphenus Seiffert et al., 2007	Tenrecoidea? (= Afrosoricida?)	Fayum, Egypt; Quarry E, Jebel Qatrani Fm	early Oligocene; Rupelian
Oligopithecus savagei Simons, 1962	Oligopithecidae; Catarrhini, Simiiformes, Anthropoidea; Primates	Fayum, Egypt; Quarry E, Jebel Qatrani Fm	early Oligocene; Rupelian
Qatrania wingi Simons and Kay, 1983	Qatraniinae, Parapithecidae; Parapithecoidea, Simiiformes, Anthropoidea; Primates	Fayum, Egypt; Quarry E, Jebel Qatrani Fm	early Oligocene; Rupelian
Indet.	Omomyiformes or Simiiformes; Primates	Fayum, Egypt; Quarry E, Jebel Qatrani Fm	early Oligocene; Rupelian
Masrasector aegypticum Simons and Gingerich, 1974	Teratodontinae, Hyainailouridae, Hyainailouroidea; Hyaenodonta	Fayum, Egypt; Quarry G, Jebel Qatrani Fm	early Oligocene; Rupelian

(continued)

Genus and species Author, year	Suprageneric taxonomy	Locality; stratigraphic level	Epoch; age
Indet.	Hyainailouridae; Hyaenodonta	Fayum, Egypt; Quarry G, Jebel Qatrani Fm	early Oligocene; Rupelian
Phiomys paraphiomyoides Wood, 1968	Phiomyidae, Hystricognathi; Rodentia	Fayum, Egypt; Quarry G, Jebel Qatrani Fm	early Oligocene; Rupelian
Metaphiomys schaubi Wood, 1968	Metaphiomyinae, Hystricognathi; Rodentia	Fayum, Egypt; Quarry G and Lower Sequence, Jebel Qatrani Fm	early Oligocene; Rupelian
Cleopatrodon robusta Bown and Simons, 1987	Ptolemaiidae; Ptolemaiida	Fayum, Egypt; Quarry I, Jebel Qatrani Fm	early Oligocene; Rupelian
Metasinopa fraasi Osborn, 1909	Teratodontinae, Hyainailouridae, Hyainailouroidea; Hyaenodonta	Fayum, Egypt; Quarry I, Jebel Qatrani Fm	early Oligocene; Rupelian
Qatranilestes oligocaenus Seiffert, 2010	Tenrecoidea (= Afrosoricida)	Fayum, Egypt; Quarry I, Jebel Qatrani Fm	early Oligocene; Rupelian
Phiocricetomys minutus Wood, 1968	Phiocricetomyinae, Hystricognathi; Rodentia	Fayum, Egypt; Quarry I, Jebel Qatrani Fm	early Oligocene; Rupelian
Monamys simonsi (Wood 1968) Sallam and Seiffert, 2020	Metaphiomyinae, Hystricognathi; Rodentia	Fayum, Egypt; Quarry I, Jebel Qatrani Fm	early Oligocene; Rupelian
cf. *Oligopithecus* sp.	Oligopithecidae, Catarrhini, Simiiformes, Anthropoidea; Primates	Fayum, Egypt; Quarry I, Jebel Qatrani Fm	early Oligocene; Rupelian
Phasmatonycteris phiomensis Gunnell et al., 2014	Myzopodidae, Noctilionoidea, Microchiroptera; Chiroptera	Fayum, Egypt; Quarry I, Jebel Qatrani Fm	early Oligocene; Rupelian
Philisis sphingis Sigé, 1985	Philisidae, Vespertilionoidea, Microchiroptera; Chiroptera	Fayum, Egypt; Quarry I, Jebel Qatrani Fm	early Oligocene; Rupelian
Qatranitherium africanum (Simons and Bown, 1984) Crochet et al., 1992	Herpetotheriidae, Didelphimorphia; Marsupialia	Fayum, Egypt; Quarry M, Jebel Qatrani Fm	early Oligocene; Rupelian

Taxon	Family; Order	Locality; Formation	Age
Metoldobotes cf. *stromeri* Schlosser, 1910	Metoldobotinae, Macroscelididae; Macroscelidea	Fayum, Egypt; Quarry M, Jebel Qatrani Fm	early Oligocene; Rupelian
Qatrania fleaglei Simons and Kay, 1988	Qatraniinae, Parapithecidae, Parapithecoidea, Simiiformes, Anthropoidea; Primates	Fayum, Egypt; Quarry M, Jebel Qatrani Fm	early Oligocene; Rupelian
Indet.	Manidae; Pholidota	Fayum, Egypt; Quarry M and L-12, Jebel Qatrani Fm	early Oligocene; Rupelian
Eosiren imenti Domning et al., 1994	Dugongidae, Halitheriinae; Sirenia	Fayum, Egypt; Quarry O, Jebel Qatrani Fm	early Oligocene; Rupelian
Cleopatrodon ayeshae Bown and Simons, 1987	Ptolemaiidae; Ptolemaiida	Fayum, Egypt; Quarry V, Jebel Qatrani Fm	early Oligocene; Rupelian
Ptolemaia grangeri Bown and Simons, 1987	Ptolemaiidae; Ptolemaiida	Fayum, Egypt; Quarry V, Jebel Qatrani Fm	early Oligocene; Rupelian
Selenohyrax chatrathi Rasmussen and Simons, 1988	Saghatheriidae; Hyracoidea	Fayum, Egypt; Quarry V, Jebel Qatrani Fm	early Oligocene; Rupelian
Saghatherium humarum Rasmussen and Simons, 1988	Saghatheriidae; Hyracoidea	Fayum, Egypt; Quarry V, Jebel Qatrani Fm	early Oligocene; Rupelian
Geniohyus magnus Andrews, 1904	Geniohyidae; Hyracoidea	Fayum, Egypt; Quarry V, Jebel Qatrani Fm	early Oligocene; Rupelian
Saghatherium antiquum Andrews and Beadnell, 1902	Saghatheriidae; Hyracoidea	Jebel al Hasawnah, Fezzan, Libya; Tarab Fm	early Oligocene; Rupelian
Phiomia serridens Andrews and Beadnell, 1902	Palaeomastodontidae, Elephantiformes; Proboscidea	Jebel Bon Gobrine, Tunisia; Grès de Fortuna Fm	early Oligocene; Rupelian
Indet.	Anthracotheriidae?	Jebel Bon Gobrine, Tunisia; Grès de Fortuna Fm	early Oligocene; Rupelian
Bunohyrax aff. *fajumensis*	Geniohyidae; Hyracoidea	Malembe, Angola	early Oligocene; Rupelian

(continued)

Genus and species Author, year	Suprageneric taxonomy	Locality; stratigraphic level	Epoch; age
Geniohyus aff. *mirus*	Geniohyidae; Hyracoidea	Malembe, Angola	early Oligocene; Rupelian
Arsinoitherium sp.	Arsinoitheriidae; Embrithopoda	Malembe, Angola	early Oligocene; Rupelian
Phiomia? sp.	Palaeomastodontidae, Elephantiformes; Proboscidea	Malembe, Angola	early Oligocene; Rupelian
Indet.	Sirenia	Malembe, Angola	early Oligocene; Rupelian
Indet.	Catarrhini, Simiiformes, Anthropoidea; Primates	Malembe, Angola	early Oligocene; Rupelian
Phiomia aff. *serridens*	Palaeomastodontidae, Elephantiformes; Proboscidea	Oued Bazina, Tunisia	early Oligocene; Rupelian
Indet.	Palaeomastodontidae, Elephantiformes; Proboscidea	Oued Bazina, Tunisia	early Oligocene; Rupelian
Qatranitherium aff. *africanum* (Simons and Bown, 1984) Crochet et al., 1992	Herpetotheriidae, Didelphimorphia; Marsupialia	Taqah, Oman, Arabian Peninsula; Ashawq Fm, Shizar Member	early Oligocene; Rupelian
Masrasector ligabuei Crochet et al., 1990	Teratodontinae, Hyainailouridae, Hyainailouroidea; Hyaenodonta	Taqah, Oman, Arabian Peninsula; Ashawq Fm, Shizar Member	early Oligocene; Rupelian
Thyrohyrax meyeri Rasmussen and Simons, 1991	Saghatheriidae; Hyracoidea	Taqah, Oman, Arabian Peninsula; Ashawq Fm, Shizar Member	early Oligocene; Rupelian
cf. *Antilohyrax* sp.	Hyracoidea	Taqah, Oman, Arabian Peninsula; Ashawq Fm, Shizar Member	early Oligocene; Rupelian
Arsinoitherium andrewsi Lankester, 1903	Arsinoitheriidae; Embrithopoda	Taqah, Oman, Arabian Peninsula; Ashawq Fm, Shizar Member	early Oligocene; Rupelian
Phiomia sp.	Palaeomastodontidae, Elephantiformes; Proboscidea	Taqah, Oman, Arabian Peninsula; Ashawq Fm, Shizar Member	early Oligocene; Rupelian
Indet.	Anomaluridae, Anomaluroidea; Rodentia	Taqah, Oman, Arabian Peninsula; Ashawq Fm, Shizar Member	early Oligocene; Rupelian

Phiomys cf. *lavocati*	Phiomyidae, Hystricognathi; Rodentia	Taqah, Oman, Arabian Peninsula; Ashawq Fm, Shizar Member	early Oligocene; Rupelian
cf. *Metaphiomys* sp.	Metaphiomyinae, Hystricognathi; Rodentia	Taqah, Oman, Arabian Peninsula; Ashawq Fm, Shizar Member	early Oligocene; Rupelian
Indet.	Cricetidae, Murida; Rodentia	Taqah, Oman, Arabian Peninsula; Ashawq Fm, Shizar Member	early Oligocene; Rupelian
"*Moeripithecus markgrafi*" Schlosser, 1910	Propliopithecidae; Catarrhini, Simiiformes, Anthropoidea; Primates	Taqah, Oman, Arabian Peninsula; Ashawq Fm, Shizar Member	early Oligocene; Rupelian
Oligopithecus rogeri Gheerbrant et al., 1995	Oligopithecidae; Catarrhini, Simiiformes, Anthropoidea; Primates	Taqah, Oman, Arabian Peninsula; Ashawq Fm, Shizar Member	early Oligocene; Rupelian
Propliopithecus sp.	Propliopithecidae; Catarrhini, Simiiformes, Anthropoidea; Primates	Taqah, Oman, Arabian Peninsula; Ashawq Fm, Shizar Member	early Oligocene; Rupelian
Shizarodon dhofarensis Gheerbrant et al., 1993	Galagidae?, Lorisiformes, Strepsirrhini; Primates	Taqah, Oman, Arabian Peninsula; Ashawq Fm, Shizar Member	early Oligocene; Rupelian
Omanodon minor Gheerbrant et al., 1993	Galagidae?, Lorisiformes, Strepsirrhini; Primates	Taqah, Oman, Arabian Peninsula; Ashawq Fm, Shizar Member	early Oligocene; Rupelian
Indet.	Microchiroptera; Chiroptera	Taqah, Oman, Arabian Peninsula; Ashawq Fm, Shizar Member	early Oligocene; Rupelian
Hipposideros (*Brachipposideros*) *omani* Sigé, 1994	Hipposideridae; Emballonuroidea, Microchiroptera; Chiroptera	Taqah, Oman, Arabian Peninsula; Ashawq Fm, Shizar Member	early Oligocene; Rupelian
Dhofarella thaleri Sigé, 1994	Emballonuridae, Emballonuroidea, Microchiroptera; Chiroptera	Taqah, Oman, Arabian Peninsula; Ashawq Fm, Shizar Member	early Oligocene; Rupelian
Chibanycteris herberti Sigé, 1994	Nycteridae, Emballonuroidea, Microchiroptera; Chiroptera	Taqah, Oman, Arabian Peninsula; Ashawq Fm, Shizar Member	early Oligocene; Rupelian

(continued)

Genus and species Author, year	Suprageneric taxonomy	Locality; stratigraphic level	Epoch; age
Indet.	Hipposideridae, Rhinolophoidea, Microchiroptera; Chiroptera	Taqah, Oman, Arabian Peninsula; Ashawq Fm, Shizar Member	early Oligocene; Rupelian
Indet.	Vespertilionoidea, Microchiroptera; Chiroptera	Taqah, Oman, Arabian Peninsula; Ashawq Fm, Shizar Member	early Oligocene; Rupelian
Philisis sevketi Sigé, 1994	Philisidae, Vespertilionoidea, Microchiroptera; Chiroptera	Taqah, Oman, Arabian Peninsula; Ashawq Fm, Shizar Member	early Oligocene; Rupelian
cf. *Philisis* sp.	Philisidae, Vespertilionoidea, Microchiroptera; Chiroptera	Taqah, Oman, Arabian Peninsula; Ashawq Fm, Shizar Member	early Oligocene; Rupelian
Indet.	Anthracotheriidae, Hippopotamoidea; Artiodactyla	Taqah, Oman, Arabian Peninsula; Ashawq Fm, Shizar Member	early Oligocene; Rupelian
cf. *Saghatherium bowni*	Saghatheriidae; Hyracoidea	Thaytiniti, Oman, Arabian Peninsula; Ashawq Fm, Shizar Member	early Oligocene; Rupelian
cf. *Thyrohyrax meyeri*	Saghatheriidae; Hyracoidea	Thaytiniti, Oman, Arabian Peninsula; Ashawq Fm, Shizar Member	early Oligocene; Rupelian
cf. *Bunohyrax* sp.	Geniohyidae; Hyracoidea	Thaytiniti, Oman, Arabian Peninsula; Ashawq Fm, Shizar Member	early Oligocene; Rupelian
Arsinoitherium andrewsi Lankester, 1903	Arsinoitheriidae; Embrithopoda	Thaytiniti, Oman, Arabian Peninsula; Ashawq Fm, Shizar Member	early Oligocene; Rupelian
Omanitherium dhofarensis Seiffert et al., 2012	Barytheriidae or Numidotheriidae; Proboscidea	Thaytiniti, Oman, Arabian Peninsula; Ashawq Fm, Shizar Member	early Oligocene; Rupelian
Indet.	Barytheriidae?; Proboscidea	Thaytiniti, Oman, Arabian Peninsula; Ashawq Fm, Shizar Member	early Oligocene; Rupelian
Phiomia or *Palaeomastodon* sp.	Palaeomastodontidae, Elephantiformes; Proboscidea	Thaytiniti, Oman, Arabian Peninsula; Ashawq Fm, Shizar Member	early Oligocene; Rupelian

Indet. (spp.)	Rodentia	Thaytiniti, Oman, Arabian Peninsula; Ashawq Fm, Shizar Member	early Oligocene; Rupelian
cf. *Metaphiomys* sp. 1	Metaphiomyinae, Hystricognathi; Rodentia	Thaytiniti, Oman, Arabian Peninsula; Ashawq Fm, Shizar Member	early Oligocene; Rupelian
cf. *Metaphiomys* sp. 2	Metaphiomyinae, Hystricognathi; Rodentia	Thaytiniti, Oman, Arabian Peninsula; Ashawq Fm, Shizar Member	early Oligocene; Rupelian
n. g., n. sp.	Oligopithecidae; Catarrhini, Simiiformes, Anthropoidea; Primates	Thaytiniti, Oman, Arabian Peninsula; Ashawq Fm, Shizar Member	early Oligocene; Rupelian
Bothriogenys gorringei Andrews and Beadnell, 1902	Anthracotheriidae, Hippopotamoidea; Artiodactyla	Thaytiniti, Oman, Arabian Peninsula; Ashawq Fm, Shizar Member	early Oligocene; Rupelian
Titanohyrax angustidens?	Titanohyracidae; Hyracoidea	Zellah, Sirt Basin, Libya; loc. 1: N 28°29′30″, E 17°36′27″, CTMD	early Oligocene; Rupelian
Prozenkerella saharaensis Coster et al., 2015	Zenkerellidae, Anomaluroidea; Rodentia	Zellah, Sirt Basin, Libya; Zellah7 Incision local fauna, CTMD	early Oligocene; Rupelian
Gaudeamus lavocati Coster et al., 2010	Gaudeamuridae, Hystricognathi; Rodentia	Zellah, Sirt Basin, Libya; Zellah loc. 5 Incision local fauna, CTMD	early Oligocene; Rupelian
Metaphiomys schaubi Wood, 1968	Metaphiomyinae, Hystricognathi; Rodentia	Zellah, Sirt Basin, Libya; Fejfar locality and Z7I locality, CTMD	early Oligocene; Rupelian
Bothriogenys cf. *gorringei*	Anthracotheriidae, Hippopotamoidea; Artiodactyla	Zellah, Sirt Basin, Libya; loc. 1. E, CTMD	early Oligocene; Rupelian
Palaeomastodon beadnelli Andrews, 1901	Palaeomastodontidae, Elephantiformes; Proboscidea	Zellah, Sirt Basin, Libya; loc. 1, CTMD	early Oligocene; Rupelian
Phiomia serridens Andrews and Beadnell, 1902	Palaeomastodontidae, Elephantiformes; Proboscidea	Zellah, Sirt Basin, Libya; loc. 1, CTMD	early Oligocene; Rupelian

(continued)

Genus and species Author, year	Suprageneric taxonomy	Locality; stratigraphic level	Epoch; age
Indet.	Sirenia	Zellah, Sirt Basin, Libya; loc. 2, Wadi Umm al Laban, CTMD	early Oligocene; Rupelian
Phiomys andrewsi Osborn, 1908	Phiomyidae, Hystricognathi; Rodentia	Zellah, Sirt Basin, Libya; loc. 2, Wadi Umm al Laban, CTMD	early Oligocene; Rupelian
Neophiomys dawsonae Coster et al., 2015	Phiocricetomyinae, Hystricognathi; Rodentia	Zellah, Sirt Basin, Libya; loc. 2, Wadi Umm al Laban, CTMD	early Oligocene; Rupelian
Metaphiomys zallahensis Coster et al., 2015	Metaphiomyinae, Hystricognathi; Rodentia	Zellah, Sirt Basin, Libya; loc. 2, Wadi Umm al Laban, CTMD	early Oligocene; Rupelian
Bothriogenys n. sp.	Anthracotheriidae; Artiodactyla	Zellah, Sirt Basin, Libya; loc. 2, Wadi Umm al Laban, CTMD	early Oligocene; Rupelian
Thyrohyrax libycus Coster et al., 2015	Saghatheriidae; Hyracoidea	Zellah, Sirt Basin, Libya; Z7I locality, CTMD	early Oligocene; Rupelian
Neophiomys paraphiomyoides (Wood, 1968) Coster et al., 2012	Thryonomyidae, Hystricognathi; Rodentia	Zellah, Sirt Basin, Libya; Z7I locality, CTMD	early Oligocene; Rupelian
Apidium zuetina Beard et al., 2016	Parapithecidae; Catarrhini, Simiiformes, Anthropoidea; Primates	Zellah, Sirt Basin, Libya; Zellah 7 Incision local fauna, CTMD	early Oligocene; Rupelian
Simonsius harujensis Mattingly et al., 2021	Parapithecinae, Parapithecidae, Catarrhini, Simiiformes, Anthropoidea; Primates	Zellah, Sirt Basin, Libya; Zellah 7 Incision local fauna, CTMD	early Oligocene; Rupelian
Africtis sirtensis Mattingly et al., 2020	Carnivoramorpha	Zellah, Sirt Basin, Libya; Zellah 7 Incision local fauna, CTMD	early Oligocene; Rupelian
Phiocricetomys atavus Coster et al., 2012	Phiocricetomyinae, Hystricognathi; Rodentia	Zellah, Sirt Basin, Libya; Incision local fauna, CTMD	early Oligocene; Rupelian
Masrasector? sp.	Teratodontinae, Hyainailouridae, Hyainailouroidea; Hyaenodonta	Lokone, Lokichar Basin, Kenya; Lokone Sandstone Fm	early or late Oligocene; Rupelian or Chattian

Taxon	Family; Order	Locality	Age
Indet.	Saghatheriidae; Hyracoidea	Lokone, Lokichar Basin, Kenya; Lokone Sandstone Fm	early or late Oligocene; Rupelian or Chattian
Phiomia sp.	Palaeomastodontidae, Elephantiformes; Proboscidea	Lokone, Lokichar Basin, Kenya; Lokone Sandstone Fm	early or late Oligocene; Rupelian or Chattian
Metaphiomys cf. *schaubi*	Metaphiomyinae, Hystricognathi; Rodentia	Lokone, Lokichar Basin, Kenya; Lokone Sandstone Fm	early or late Oligocene; Rupelian or Chattian
Turkanamys hexalophus Marivaux et al., 2012	Phiomyidae, Hystricognathi; Rodentia	Lokone, Lokichar Basin, Kenya; Lokone Sandstone Fm	early or late Oligocene; Rupelian or Chattian
Lokonepithecus manai Ducrocq et al., 2011	Parapithecidae; Primates, Anthropoidea	Lokone, Lokichar Basin, Kenya; Lokone Sandstone Fm	early or late Oligocene; Rupelian or Chattian
Epirigenys lokonensis Lihoreau et al., 2015	Anthracotheriidae, Hippopotamoidea; Artiodactyla	Lokone, Lokichar Basin, Kenya; Lokone Sandstone Fm	early or late Oligocene; Rupelian or Chattian
Apterodon aff. *gaudryi*	Apterodontinae, Hyainailouridae, Hyainailouroidea; Hyaenodonta	Lokone, Lokichar Basin, Kenya; Lokone Sandstone Fm	early or late Oligocene; Rupelian or Chattian
Prodeinotherium cf. *hobleyi*	Deinotheriidae, Deinotherioidea; Proboscidea	Adi Ugri, Eritrea; Intertrappean Beds, K/Ar [39/40], 23.5 Ma	latest Oligocene; Chattian
n. g., n. sp.	Hyracoidea	Chilga, Ethiopia; 28–27 Ma (Kappelman et al., 2003)	late Oligocene; Chattian
Pachyhyrax n. sp.	Geniohyidae; Hyracoidea	Chilga, Ethiopia; 28–27 Ma (Kappelman et al., 2003)	late Oligocene; Chattian
Bunohyrax sp.	Geniohyidae; Hyracoidea	Chilga, Ethiopia; 28–27 Ma (Kappelman et al., 2003)	late Oligocene; Chattian
Arsinoitherium giganteum Sanders et al., 2004	Arsinoitheriidae; Embrithopoda	Chilga, Ethiopia; 28–27 Ma (Kappelman et al., 2003)	late Oligocene; Chattian
Chilgatherium harrisi Sanders et al., 2004	Deinotheriidae, Deinotherioidea; Proboscidea	Chilga, Ethiopia; 28–27 Ma (Kappelman et al., 2003)	late Oligocene; Chattian

(*continued*)

Genus and species Author, year	Suprageneric taxonomy	Locality; stratigraphic level	Epoch; age
Phiomia major Sanders et al., 2004	Palaeomastodontidae, Elephantiformes; Proboscidea	Chilga, Ethiopia; 28–27 Ma (Kappelman et al., 2003)	late Oligocene; Chattian
Palaeomastodon n. sp. 1	Palaeomastodontidae, Elephantiformes; Proboscidea	Chilga, Ethiopia; 28–27 Ma (Kappelman et al., 2003)	late Oligocene; Chattian
Palaeomastodon n. sp. 2	Palaeomastodontidae, Elephantiformes; Proboscidea	Chilga, Ethiopia; 28–27 Ma (Kappelman et al., 2003)	late Oligocene; Chattian
cf. *Gomphotherium* n. sp.	Gomphotheriidae, Elephantida, Elephantimorpha, Elephantiformes; Proboscidea,	Chilga, Ethiopia; 28–27 Ma (Kappelman et al., 2003)	late Oligocene; Chattian
Eritreum melakeghebrekristosi Shoshani et al., 2006	Elephantimorpha, Elephantiformes; Proboscidea	Dogali, Eritrea; ^{40}Ar/^{39}Ar dating, 26.8 ± 1.5 Ma	late Oligocene; Chattian
Megalohyrax eocaenus Andrews, 1903	Geniohyidae; Hyracoidea	Harrat Al Ujayfa, W Saudi Arabia; Middle Shumaysi Fm, 29–28 Ma	late Oligocene; Chattian
Geniohyus or *Bunohyrax* sp.	Geniohyidae; Hyracoidea	Harrat Al Ujayfa, W Saudi Arabia; Middle Shumaysi Fm, 28–29 Ma	late Oligocene; Chattian
Arsinoitherium cf. *zitelli*	Arsinoitheriidae; Embrithopoda	Harrat Al Ujayfa, W Saudi Arabia; Middle Shumaysi Fm, 28–29 Ma	late Oligocene; Chattian
Palaeomastodon sp.	Palaeomastodontidae, Elephantiformes; Proboscidea	Harrat Al Ujayfa, W Saudi Arabia; Middle Shumaysi Fm, 28–29 Ma	late Oligocene; Chattian
cf. *Gomphotherium* sp.	Gomphotheriidae, Elephantida, Elephantimorpha, Elephantiformes; Proboscidea	Harrat Al Ujayfa, W Saudi Arabia; Middle Shumaysi Fm, 28–29 Ma	late Oligocene; Chattian
Indet.	Mammutidae, Elephantida, Elephantimorpha, Elephantiformes; Proboscidea	Harrat Al Ujayfa, W Saudi Arabia; Middle Shumaysi Fm, 28–29 Ma	late Oligocene; Chattian

Saadanius hijazensis Zalmout et al., 2010	Saadaniidae, Catarrhini, Anthropoidea; Primates	Harrat Al Ujayfa, W Saudi Arabia; Middle Shumaysi Fm, 28–29 Ma	late Oligocene; Chattian
Bothriogenys fraasi Schmidt, 1913	Anthracotheriidae; Artiodactyla	Harrat Al Ujayfa, W Saudi Arabia; Middle Shumaysi Fm, 28–29 Ma	late Oligocene; Chattian
Hyainailouros sp.	Hyainailourinae, Hyainailouridae, Hyainailouroidea; Hyaenodonta	Losodok (= Lothidok), Kenya; Eragaleit Beds (Intertrappean), K/Ar 25–24 Ma	late Oligocene; Chattian
Mlanyama sugu Rasmussen and Gutierrez, 2009	Hyainailourinae, Hyainailouridae, Hyainailouroidea; Hyaenodonta	Losodok (= Lothidok), Kenya; Eragaleit Beds (Intertrappean), K/Ar 25–24 Ma	late Oligocene; Chattian
Thyrohyrax kenyaensis Rasmussen and Gutierrez, 2009	Saghatheriidae; Hyracoidea	Losodok (= Lothidok), Kenya; Eragaleit Beds (Intertrappean), K/Ar 25–24 Ma	late Oligocene; Chattian
Afrohyrax sp.	Pliohyracidae; Hyracoidea	Losodok (= Lothidok), Kenya; Eragaleit Beds (Intertrappean), K/Ar 25–24 Ma	late Oligocene; Chattian
Thyrohyrax microdon Rasmussen and Gutierrez, 2009	Saghatheriidae; Hyracoidea	Losodok (= Lothidok), Kenya; Eragaleit Beds (Intertrappean), K/Ar 25–24 Ma	late Oligocene; Chattian
Brachyhyrax oligocenus Rasmussen and Gutierrez, 2009	Geniohyidae; Hyracoidea	Losodok (= Lothidok), Kenya; Eragaleit Beds (Intertrappean), K/Ar 25–24 Ma	late Oligocene; Chattian
Meroehyrax kyongoi Rasmussen and Gutierrez, 2009	Pliohyracidae; Hyracoidea	Losodok (= Lothidok), Kenya; Eragaleit Beds (Intertrappean), K/Ar 25–24 Ma	late Oligocene; Chattian
Arsinoitherium sp.	Arsinoitheriidae; Embrithopoda	Losodok (= Lothidok), Kenya; Eragaleit Beds (Intertrappean), K/Ar 25–24 Ma	late Oligocene; Chattian
Prodeinotherium? sp.	Deinotheriidae, Deinotherioidea; Proboscidea	Losodok (= Lothidok), Kenya; Eragaleit Beds (Intertrappean), K/Ar 25–24 Ma	late Oligocene; Chattian
Losodokodon losodokius Rasmussen and Gutierrez, 2009	Mammutidae, Mammutida, Elephantimorpha, Elephantiformes; Proboscidea	Losodok (= Lothidok), Kenya; Eragaleit Beds (Intertrappean), K/Ar 25–24 Ma	late Oligocene; Chattian

(continued)

Genus and species Author, year	Suprageneric taxonomy	Locality; stratigraphic level	Epoch; age
Diamantomys timius Rasmussen and Gutierrez, 2009	Diamantomyidae, Hystricognathi; Rodentia	Losodok (= Lothidok), Kenya; Eragaleit Beds (Intertrappean), K/Ar 25–24 Ma	late Oligocene; Chattian
Kamoyapithecus hamiltoni (Madden, 1980) Leakey et al., 1995	Hominoidea, Catarrhini, Simiiformes, Anthropoidea; Primates	Losodok (= Lothidok), Kenya; Eragaleit Beds (Intertrappean), K/Ar 25–24 Ma	late Oligocene; Chattian
Brachyodus sp.	Anthracotheriidae; Artiodactyla	Losodok (= Lothidok), Kenya; Eragaleit Beds (Intertrappean), K/Ar 25–24 Ma	late Oligocene; Chattian
cf. *Prodeinotherium* sp.	Deinotheriidae, Deinotherioidea; Proboscidea	Mai Gobro, Eritrea; K/A 24.6 Ma	latest Oligocene; Chattian
cf. *Gomphotherium* sp.	Gomphotheriidae, Elephantida, Elephantimorpha, Elephantiformes; Proboscidea	Mai Gobro, Eritrea; K/A 24.6 Ma	latest Oligocene; Chattian
cf. *Antilohyrax* sp.	Titanohyracidae; Hyracoidea	Minqar Tibaghbagh, Qattara Depression, Egypt; "BOTM"	late Oligocene; Chattian
cf. *Phiomia* sp.	Palaeomastodontidae, Elephantiformes; Proboscidea	Minqar Tibaghbagh, Qattara Depression, Egypt; "BOTM"	late Oligocene; Chattian
Eosiren? sp.	Sirenia	Minqar Tibaghbagh, Qattara Depression, Egypt; "BOTM"	late Oligocene; Chattian
cf. *Bothriogenys* sp.	Anthracotheriidae; Artiodactyla	Minqar Tibaghbagh, Qattara Depression, Egypt; "BOTM"	late Oligocene; Chattian
Pakakali rukwaensis Borth and Stevens, 2017	Hyainailourinae, Hyainailouridae, Hyainailouroidea; Hyaenodonta	Rukwa Basin, Tanzania; Nsungwe Fm, Songwe Member, paleomagnetic zone C8n.1n, 25.2 Ma	late Oligocene; Chattian
Rukwasengi butleri Stevens et al., 2021	Myohyracinae, Macroscelididae; Macroscelidea	Rukwa Basin, Tanzania; Nsungwe Fm, Songwe Member, paleomagnetic zone C8n.1n, 25.2 Ma	late Oligocene; Chattian

Oligorhynchocyon songwensis Stevens et al., 2021	Rhynchocyoninae, Macroscelididae; Macroscelidea	Rukwa Basin, Tanzania; Nsungwe Fm, Songwe Member, paleomagnetic zone C8n.1n, 25.2 Ma	late Oligocene; Chattian
Rukwalorax jinokitana Stevens et al., 2009	Hyracoidea	Rukwa Basin, Tanzania; Nsungwe Fm, Songwe Member, paleomagnetic zone C8n.1n, 25.2 Ma	late Oligocene; Chattian
Metaphiomys cf. *beadnelli*	Metaphiomyinae, Hystricognathi; Rodentia	Rukwa Basin, Tanzania; Nsungwe Fm, Songwe Member, paleomagnetic zone C8n.1n, 25.2 Ma	late Oligocene; Chattian
Kahawamys mbeyaensis Stevens et al., 2009	Thryonomyoidea, Hystricognathi; Rodentia	Rukwa Basin, Tanzania; Nsungwe Fm, Songwe Member, paleomagnetic zone C8n.1n, 25.2 Ma	late Oligocene; Chattian
Indet.	Anthropoidea; Primates	Rukwa Basin, Tanzania; Nsungwe Fm, Songwe Member, paleomagnetic zone C8n.1n, 25.2 Ma	late Oligocene; Chattian
Rukwapithecus fleaglei Stevens et al., 2013	Hominoidea, Catarrhini, Anthropoidea; Primates	Rukwa Basin, Tanzania; Nsungwe Fm, Songwe Member, paleomagnetic zone C8n.1n, 25.2 Ma	late Oligocene; Chattian
Nsungwepithecus gunnelli Stevens et al., 2013	Cercopithecoidea, Catarrhini, Anthropoidea; Primates	Rukwa Basin, Tanzania; Nsungwe Fm, Songwe Member, paleomagnetic zone C8n.1n, 25.2 Ma	late Oligocene; Chattian
Indet.	Tenrecoidea (= Afrosoricida)	Eocliff, Sperrgebiet, Namibia	late Oligocene?; Chattian?
Namachloris arenatans Pickford, 2015	Chrysochloridae; Tenrecoidea (= Afrosoricida)	Eocliff, Sperrgebiet, Namibia	late Oligocene?; Chattian?
Eorhynchocyon rupestris Senut and Pickford, 2021	Rhynchocyonidae; Macroscelidea	Eocliff, Sperrgebiet, Namibia	late Oligocene?; Chattian?

(continued)

Genus and species Author, year	Suprageneric taxonomy	Locality; stratigraphic level	Epoch; age
Namasengi mockeae Senut and Pickford, 2021	Namasenginae, Macroscelididae; Macroscelidea	Eocliff, Sperrgebiet, Namibia	late Oligocene?; Chattian?
Promyohyrax namibiensis Senut and Pickford, 2021	Myohyracidae; Macroscelidea	Eocliff, Sperrgebiet, Namibia	late Oligocene?; Chattian?
Afrohypselodontus minus Senut and Pickford, 2021	Afrohypselodontidae; Macroscelidea	Eocliff, Sperrgebiet, Namibia	late Oligocene?; Chattian?
Afrohypselodontus grandis Senut and Pickford, 2021	Afrohypselodontidae; Macroscelidea	Eocliff, Sperrgebiet, Namibia	late Oligocene?; Chattian?
Indet.	Titanohyracidae; Hyracoidea	Eocliff, Sperrgebiet, Namibia	late Oligocene?; Chattian?
Prepomonomys bogenfelsi Pickford et al., 2008	Diamantomyidae, Hystricognathi; Rodentia	Eocliff, Sperrgebiet, Namibia	late Oligocene?; Chattian?
Protophiomys cf. *algeriensis*	Protophiomyinae, Hystricognathi; Rodentia	Eocliff, Sperrgebiet, Namibia	late Oligocene?; Chattian?
Metaphiomys cf. *schaubi* Wood, 1968	Metaphiomyinae, Hystricognathi; Rodentia	Eocliff, Sperrgebiet, Namibia	late Oligocene?; Chattian?
Silicamys cingulatus Pickford et al., 2008	Myophiomyidae, Hystricognathi; Rodentia	Eocliff, Sperrgebiet, Namibia	late Oligocene?; Chattian?
Indet.	Rodentia	Eocliff, Sperrgebiet, Namibia	late Oligocene?; Chattian?
Phiomys aff. *phiomyoides*	Phiomyidae, Hystricognathi; Rodentia	Eocliff, Sperrgebiet, Namibia	late Oligocene?; Chattian?
Tufamys woodi Pickford, 2018	Tufamyidae, Hystricognathi; Rodentia Thryonomyoidea	Eocliff, Sperrgebiet, Namibia	late Oligocene?; Chattian?
Namaloris rupestris Pickford, 2015	Lorisidae, Strepsirrhini; Primates, Lorisiformes	Eocliff, Sperrgebiet, Namibia	late Oligocene?; Chattian?

Arenagale calcareus Pickford, 2015	Tenrecinae; Tenrecoidea (= Afrosoricida)	EC8, Eocliff, Sperrgebiet, Namibia	late Oligocene?; Chattian?
Namagale grandis Pickford, 2015	Potamogalinae; Tenrecoidea (= Afrosoricida)	EC8, Eocliff, Sperrgebiet, Namibia	late Oligocene?; Chattian?
Sperrgale minutus Pickford, 2015	Tenrecinae; Tenrecoidea (= Afrosoricida)	EC8, Eocliff, Sperrgebiet, Namibia	late Oligocene?; Chattian?
Phiomys aff. *lavocati*	Phiomyidae, Hystricognathi; Rodentia	EC8, Eocliff, Sperrgebiet, Namibia	late Oligocene?; Chattian?
Sperrgale sp.	Tenrecinae; Tenrecoidea (= Afrosoricida)	Sperrgebiet, Namibia; Eoridge	late Oligocene?; Chattian?
Rupestrohyrax palustris Pickford, 2015	Titanohyracidae; Hyracoidea	Sperrgebiet, Namibia; Eoridge	late Oligocene?; Chattian?
Silicamys sp.	Myophiomyidae, Hystricognathi; Rodentia	Sperrgebiet, Namibia; Eoridge	late Oligocene?; Chattian?
Bothriogenys gorringei (Andrews and Beadnell, 1902)	Anthracotheriidae; Artiodactyla	Sperrgebiet, Namibia; Eoridge	late Oligocene?; Chattian?
Silicamys cingulatus Pickford et al., 2008	Myophiomyidae, Hystricognathi; Rodentia	Silica North, Sperrgebiet, Namibia	late Oligocene?; Chattian?
Tufamys woodi Pickford, 2018	Tufamyidae, Hystricognathi; Rodentia	Silica North, Sperrgebiet, Namibia	late Oligocene?; Chattian?
Indet.	Catarrhini, Simiiformes, Anthropoidea; Primates	Silica North, Sperrgebiet, Namibia	late Oligocene?; Chattian?
Indet.	Teratodontinae, Hyainailouridae, Hyainailouroidea; Hyaenodonta	Silica North, Sperrgebiet, Namibia and South	late Oligocene?; Chattian?
Prepomonomys bogenfelsi Pickford et al., 2008	Diamantomyidae, Hystricognathi; Rodentia	Silica South, Sperrgebiet, Namibia	late Oligocene?; Chattian?
Protophiomys cf. *algeriensis* Jaeger et al., 1985	Protophiomyinae, Hystricognathi; Rodentia	Silica South, Sperrgebiet, Namibia	late Oligocene?; Chattian?

Glossary

adaptive radiation. *See* evolutionary radiation.

African faunal strata. The faunal assemblages that succeeded in Africa. The Paleogene mammals known in Africa belong to several distinct African faunal strata comprising different lineages (clades) originating from allochthonous stem groups that arrived in Africa during different periods of dispersal. The immigrant stem groups gave rise to endemic lineages, such as anomaluroid rodents and anthropoid primates, which evolved in Africa. The extant native mammals of Africa, such as elephants, dassies, otter shrews (potamogales), anomalures, mole-rats, springhares, galagos, monkeys, and great apes, evolved from the successive ancient African faunal strata. Currently six Cretaceous–Paleogene African faunal strata of placental mammals have been identified.

Afrosoricida. *See* Tenrecoidea.

Amphibious. Synonymous with semiaquatic. Adapted for life in both aquatic and terrestrial environments. Refers, in particular, to mammals that feed in an aquatic environment, such as sea cows and some proboscideans (moeritheres) that ate aquatic plants.

ancestral character. Also called primitive character or plesiomorphic character. A generalized character that is present in stem groups but lost (transformed) in specialized or advanced groups. For example, the tritubercular (or tribosphenic) upper molar (molar with three main tubercles) is ancestral in placentals, whereas the quadritubercular upper molar (molar with four main tubercles) is a derived character of ungulate-like lineages among placentals.

anomaluroids. Specialized rodents that have evolved in Africa since the Eocene (from 48 Ma) and include the extant flying squirrels (anomalures).

Arabo-African plate. The former great landmass that comprised Africa and the Arabian Peninsula before the opening of the Red Sea during the Oligocene–Miocene transition (23 Ma). It was isolated from the other continents as Island Africa from the mid-Cretaceous (110–100 Ma) to the Oligocene–Miocene transition (23 Ma).

arboreal. Relating to an animal adapted to life in the trees, such as a monkey or squirrel.

basal. An ancestral or stem branch/group. This is a taxon that is related the root of other relative taxa. A stem or basal group is the sister group to other taxa from the clade it belongs.

bilophodont teeth. Lophodont teeth that have only two lophs (i.e., two long, transverse, specialized cutting crests). This is the ancestral plan of the lophodont dental morphology in mammals. From this initial lophodont stage with two lophs, several lineages, such as the order of elephants, have further specialized by the development of additional lophs (e.g., trilophodonty or polylophodonty). *See also* lophodont teeth.

Boreoeutheria. One of the three main placental clades. This is the group of placentals that originated in northern continents (Boreal continents). It includes Euarchontoglires (primates, colugos, rodents) and Laurasiatheria (carnivorans, bats, true insectivores, euungulates, etc.).

bunodont teeth. Specialized teeth with low, rounded, robust reliefs (crests and tubercles) and crowns. They have a primarily grinding-crushing function and are found in mammals with a frugivorous, graminivorous, or omnivorous diet.

carnivorous. Feeding on meat. Carnivores, such as felids, specialize in eating meat.

catarrhines. The Old World anthropoid primates that include Old World monkeys (cercopithecoids), apes (hominoids), and humans (hominids). They are distinguished by several traits from South American anthropoids (platyrrhines), such as the loss of a premolar in each half of the upper or lower jaws (i.e., in the dental formula).

Caviomorpha. South American rodents, such as porcupines, agoutis, chinchillas, and capybaras, belonging to the Hystricognathi group, which also includes African rodents of the Phiomorpha group (cane rats).

Cenozoic. Synonymous with Tertiary. The geological era extending between 66 Ma and the present (the last 66 million years). It corresponds to the "Age of Mammals."

crown group. A modern group that includes all living species of that group and the extinct relatives of all lineages included in it. Placentalia is a crown group of modern mammals, whereas Eutheria includes both the stem (or basal) placentals and the crown placentals.

cursorial. Specialized locomotion for running, such as in most ungulate placentals.

dental formula. Also called a tooth formula. The number of teeth in each half of the upper or lower jaws. The ancestral dental formula of all placentals is 3/3, 1/1, 4/4, 3/3: This means the presence of three upper and lower incisors, one upper and lower canine, four upper and lower premolars, and three upper and lower molars on each side of the upper and lower jaw (44 teeth total). In the course of evolution, many placental lineages, especially the ungulates, reduced their number of teeth (e.g., with loss of canines and anterior premolars).

derived character. Also called specialized character or apomorphic character. An advanced trait that has evolved as a specialization in a particular group and is distinctive with respect to other related taxa. For instance, the carnassial teeth of the Carnivora are derived and distinctive with respect to the teeth of other placentals. Derived characters can be unique to a group, or they could have evolved by convergence in several groups as with, for example, the bilophodont teeth, which appeared both in the paenungulates and the euungulates (perissodactyls).

digitigrade. Posture of the limbs on the ground related to cursorial locomotion, in which the animal's body rests only on the extremities of the hand and foot, such as the digits and part of the metapodials (metacarpals and metatarsals), with the heel and wrist lifted off the ground. Carnivorous mammals such as cats and dogs are digitigrade.

ectoloph. The outer cutting crest, buccal (toward the cheek), of upper molars and premolars that extends longitudinally (not transversely like the lophs of bilophodont teeth).

endemic. Synonymous with native. A group of plants or animals the geographic distribution of which is restricted to the area from which it originated. An endemic species belongs to a clade (lineage) generally distinctive in morphology. Afrotheria, and in particular hyracoids, are endemic to Africa in the Paleogene times.

Euarchontoglires. A clade of placental mammals belonging to Boreoeutheria and including rodents, rabbits, tree shrews, colugos, and primates. *See also* Boreoeutheria.

Eutheria, eutherians. The group of mammals that includes stem and living placentals. Eutheria and Metatheria (marsupials) form the modern mammal group Theria (mammals with tritubercular molars; *see* tritubercular/tribosphenic teeth).

evolutionary radiation. A rapid diversification of groups marked by high rates of evolution and usually associated with an increase in morphological disparity and adaptive specializations. When a rapid diversification occurs in the same lineage—for example, proboscideans, hyracoids, anthropoids, or rodents—this is referred to as an adaptive radiation.

explosive model. One of the three evolutionary models explaining the origin and initial diversification (radiation) of placental mammals. It posits that the origin and diversification of modern orders of placentals occurred rapidly at the beginning of the Cenozoic. Paleontology best supports an explosive model of the evolutionary radiation of placental mammals. There are two variants of the explosive model of the initial diversification of modern mammals: The "hard" explosive model proposes that the initial diversification of placentals began only at the beginning of the Cenozoic, whereas the "soft" explosive model proposes that some major placentals clades (superorders) such as Afrotheria, Laurasiatheria, and Euarchontoglires originated before the Cenozoic, during the Late Cretaceous, and were at the root of the rapid early Cenozoic radiation of placentals (e.g., orders and families). Compare short fuse model and long fuse model.

folivorous. Feeding on leaves. Folivores, such as tapirs, colobine monkeys, and okapis, specialize in eating leaves.

frugivorous. Feeding on fruit. Frugivores, such as orangutans, owl monkeys, and fruit bats, specialize in eating fruit.

geodispersal. A major continental-scale biotic dispersal event involving various lineages. Well-documented examples of geodispersals of continental vertebrates include the Great American Biotic Interchange between North and South America, which resulted from the formation of the Isthmus of Panama about 3 Ma, and the Great Old World Interchange (GOWI) between Africa and Eurasia, which followed the closure of the Tethys Sea (Africa-Eurasia collision) about 23 Ma.

geomolecular dating. The dating of groups by using phylogeny and molecular-clock data, as well as paleontological calibration based on oldest fossil occurrences.

ghost lineage. A lineage implied by its relationships (phylogeny) but not documented because of gaps in the known fossil record.

Gondwana. Also known as Neo-Gondwana. An ancient supercontinent that existed in the Mesozoic, from 170 Ma, that gathered all the southern continents (South America, Africa, Antarctica, Australia, Madagascar, and India) into a single continental mass located south of the Tethys Sea and Laurasia. West Gondwana included Africa and South America; it split when the South Atlantic opened in the middle Cretaceous (by 110 Ma).

graviportal. A skeletal posture of some large mammals (and other tetrapod vertebrates) in which the limbs are columnar and erect vertically below the body. The graviportal posture is a typical adaptation of large land mammals with a high body mass, such as elephants.

Great Old World Interchange (GOWI). A major continental faunal event that took place about 23 Ma. It marked the end of Africa's continental isolation (Africa-Eurasia collision), and it is illustrated by massive faunal interchanges between Eurasia and Africa, geographic expansion and new radiations of African clades, and important extinctions of the native African Paleogene mammals, such as all embrithopods, most hyaenodonts, and many hyracoids.

herbivorous. Feeding on plants. Herbivores specialize in eating plants. Frugivores and folivores are types of herbivores.

holotype. The type reference specimen on which the definition of a species is based. The scientific name of a species is based on the type specimen, or holotype.

hypercarnivorous. Specializing nearly exclusively in feeding on meat (typically >70%). Hypercarnivorous mammals have simplified sectorial teeth and sharp and large canines.

hyperdilambdodont teeth. Upper teeth in which the ectoloph (the external crests; i.e., paracone and metacone crests) has a W-like shape and is elongated transversely. In the most derived hyperdilambdodont taxa, such as *Arsinoitherium*, the ectoloph loses its W-like shape and is simplified into two long transverse lophs resembling the lophodont teeth but in a construction that is not homologous to them.

hypocone. The fourth tubercle that develops in quadritubercular upper molars; it is absent in the primitive tritubercular (and triangular) upper molars. It adds a crushing and grinding function to the teeth during mastication, and it evolved in herbivorous mammals, such as ungulate-like mammals and primates.

hypsodont teeth. Teeth that grow continuously and have a high crown.

insectivorous. Feeding on insects. Insectivorous mammals, which are specialized in eating insects, belong to diverse mammal groups, such as bats, shrews, hedgehogs, potamogales, and elephant shrews. True insectivores belong the Eulipotyphla order (Laurasiatheria). African insectivores belong to the Afrosoricida (Afrotheria).

Island Africa. The vast continental landmass, including the African and Arabian plates, isolated from other continents for some 80 million years, from the middle Cretaceous (110–100 Ma) to the Oligocene–Miocene boundary (23 Ma).

KPg crisis, KPg extinction event. Also known as the KT (Cretaceous–Tertiary) crisis. A major global biotic event that occurred 66 Ma, at the Cretaceous–Paleogene boundary. It was marked by mass extinctions, including all non-avian dinosaurs, and by an evolutionary recovery that led to the emergence and explosion of modern taxa, such as birds and mammals. It is linked to major paleoenvironmental changes caused by the impact of a large asteroid (marked by the Chicxulub crater in Mexico) and a phase of massive volcanism in India.

Laurasia. An ancient supercontinent that existed between 170 and 56 Ma and that gathered all the northern continents (North America, Europe, and Asia) in a single landmass located north of the Tethys Sea and Gondwana.

laurasiatherians. One of the four major groups (clades) of placental mammals, including seven modern orders that originated in Laurasia—such as true insectivores (Eulipotyphla), bats (Chiroptera), carnivores (Carnivora), pangolins (Pholidota), even-toed ungulates (Artiodactyla), and odd-toed ungulates (Perissodactyla)— and their extinct relatives. Laurasiatheria is the sister group of the Euarchontoglires (Rodentia, Lagomorpha, Primates), with which it forms the Boreoeutheria.

long fuse model. One of three evolutionary models explaining the origin and initial diversification (radiation) of placental mammals. The long fuse model states that some modern orders have stem groups extending into the Cretaceous. Compare explosive model and short fuse model.

lophodont teeth. Teeth found in herbivorous mammals that are characterized by the presence of cutting crests, called lophs, that are elongated transversely. During mastication, the lophs function between the upper and lower teeth like the two blades of a pair of scissors and following a transverse movement of the jaws. This is a specialized morphology that evolved in mammals with a folivorous diet.

Ma. Million years ago (abbreviation for mega-annum).

marsupials. One of the three main groups of modern (extant) mammals, along with placentals and monotremes (platypuses). Marsupials include kangaroos, opossums, and koalas. They are characterized by a distinctive mode of reproduction and are known as "pouched mammals." Marsupials and their stem groups belong to the Metatheria, which is the sister group to the Eutheria (placentals and their stem groups), both of which belong to the modern Theria group (*see* Theria). Compare monotremes and placentals.

monotremes. One of three main groups of modern (extant) mammals, along with the placentals and marsupials. Monotremes are egg-laying mammals, today represented only by platypuses and echidnas from Australia and New Guinea. Compare marsupials and placentals.

nocturnal. Active at night, such as for feeding.

omnivorous. Feeding on both animals and plants.

orbit. Hollow part of the bony skull that encloses the eyeball and associated structures.

paleobiogeography. The study of the geographic distribution of fauna and flora in the past at different scales, from large continental realms to regional provinces and biotic assemblages.

Paleocene–Eocene Thermal Maximum (PETM). A short but major global climatic warming event (about 200,000 years) that occurred at the Paleocene–Eocene boundary, 56 Ma. This event saw a rapid global temperature rise of about 6°C and a rise in sea levels because of the warming of the oceans. It strongly affected the evolution of flora and fauna.

Paleogene. Geological period (system) corresponding to the first half of the Cenozoic, between 66 and 23 Ma. It includes the Paleocene (66–56 Ma), Eocene (56–34 Ma), and Oligocene (34–23 Ma) ages (series). During this period, Africa was an island continent (Island Africa), and the African mammals evolved isolated from those in other continents. This led to the rise of original (endemic) lineages unknown elsewhere, with some eventually going extinct and others representing large modern groups, such as that of the elephants.

paleogeography. The past geography of the world as reconstructed from geological and geodynamic data. This includes continental configurations completely different from the present world, with large supercontinents, such as Pangea, Laurasia, and Gondwana, and ancient seas like the Tethys, between Laurasia and Gondwana. Paleogeographic maps have reconstructed the vast Island Africa from the middle of the Cretaceous (ca. 110 Ma) to the end of the Oligocene (23 Ma).

paraphyletic. Related to an unnatural grouping of plants or animals that includes a common ancestor but not all its descendants. A group is said to be paraphyletic with respect to the excluded subgroups. Paraphyletic groups are united by shared primitive (ancestral) features instead of original (derived) features to that group.

phylogeny, phylogenetic relationships. The evolutionary relationships of animals and plants. A phylogenetic tree is inferred from the study of the distribution of the shared traits (molecular or morphological) of species, using parsimony or probabilistic methods.

placentals, Placentalia. One of the three major groups of extant mammals, along with marsupials and monotremes. Placentalia includes the majority of living mammals, including rodents, bats, primates, carnivores, whales, horses, pigs, elephants, and shrews. This group of mammals is characterized by a mode of reproduction in which the fetus develops to a well-advanced stage of development before birth. *See also* Eutheria. Compare marsupials and monotremes.

plantigrade. Posture of the limbs on the ground related to locomotion in which the body of the animal rests on the entire hand and foot (e.g., heel, sole, and toes). A plantigrade mammal is a "flat footed" mammal.

platyrrhines. The native South American monkeys, including the capuchins, spider monkeys, and tamarins. Platyrrhini is a subdivision of a more inclusive primate group called Anthropoidea, which includes the Catarrhini (i.e., the Old World monkeys, apes, and humans). Platyrrhines differ from catarrhines by several traits, such as the possession of a prehensile tail.

plesiomorphic character. *See* ancestral character.

pneumatization, pneumatized bones. Bones lightened by internal cavities. This is a feature of large mammals such as elephants.

primitive character. *See* ancestral character.

provincialism. A restriction of the geographic distribution of organisms relative to a broader continental distribution, for example. Some groups of animals and plants are confined to greater or lesser areas called faunal provinces, where they form original associations (species or group composition varies significantly in different provinces). The regional distribution of organisms is controlled by the presence of barriers related to climatic (e.g., temperature and rainfall) or geographic (e.g., mountains and seas for continental organisms) variations.

saltatorial. Related to the type of terrestrial locomotion of mammals and other vertebrates that move by hopping on the two hind legs.

screening-washing. A method of collecting small fossils, such as isolated teeth and bones of the microfauna, whereby raw sediments taken from a fossiliferous rock layer are washed through a series of sieves or screens of various mesh sizes to remove clay, silt, sand, and other minerals in order to concentrate the microfossils. Screening-washing is usually done in the field near the fossil site, while the sorting of the microfossils from the concentrate is performed under a binocular microscope in the laboratory.

selenodont teeth. Teeth of ungulate mammals in which the crests are developed and have a crescent shape (when seen from above, on their occlusal face). These teeth are present in herbivorous mammals whose chewing involves significant lateral movement, and they are well adapted for breaking up tough plant material. Many hyracoids and most artiodactyls, such as ruminants, have selenodont teeth.

sexual dimorphism. A condition where the males and females of a species differ from one another in size or morphology (other than the reproductive organs). For example, the size of canines differs considerably between males and females in monkeys and several other mammals. Sexual dimorphism is common in gregarious mammals.

short fuse model. One of three evolutionary models explaining the origin and initial diversification (radiation) of placental mammals. The short fuse model postulates an ancient origin of the modern orders of placentals going back to the beginning of the Cretaceous, with a first diversification within the orders at the end of the Cretaceous. Compare explosive model and long fuse model.

specialization, specialized character. *See* derived character.

stem group. An extinct species or group related to the root of an extant evolutionary lineage (crown group). The stem group evolved before living members of the same lineage began to diverge from each other. For example, stem placentals (basal eutherians), such as the generalized insectivores of the cimolestid family, and stem proboscideans, such as *Phosphatherium*.

stratigraphy. A branch of geology whose purpose is the identification, description, and relative dating of sedimentary beds. Together with various other data, such as those of biostratigraphy and geochronology, it is the basis for establishing the geological time scale.

talonid. The back (posterior) part of the lower tooth (usually lower molar) of a mammal with tritubercular molars. It has a primary crushing function during mastication.

Tenrecoidea, tenrecoids. Synonymous with Afrosoricida. Insectivorous afrotherian mammals that include tenrecs (Tenrecinae), otter shrews (Potamogalinae), and golden moles (Chrysochloridae). They are characterized by peculiar teeth with zalambdodont morphology (*see* zalambdodont teeth).

Tethys Sea. The ancient sea that separated Laurasia from Gondwana. It existed between about 170 and 23 Ma and disappeared with the collision of Africa and Eurasia, which marked the end of Africa's isolation and the birth of the Old World faunal realm.

Theria, therians. The crown group of modern mammals that includes all living marsupials and placentals. It has many peculiar derived morphological traits, a key one being the presence of the tribosphenic or tritubercular molars (*see* tritubercular/tribosphenic teeth). Boreosphenida includes Theria and its stem groups.

tritubercular/tribosphenic teeth. Molars with a crushing and grinding function with a mortar-and-pestle-like morphology. The tritubercular/tribosphenic upper molars have three cusps (one new) arranged in a triangle around a crushing-grinding basin (protofossa); the lower molars have a posterior expansion with a posterior crushing-grinding basin (talonid basin). The mortar is represented by the two basins, and the pestle by the new cusps (protocone of upper molars; hypoconid of lower molars). The tribosphenic molar is a major morphological and adaptive innovation of the Theria group (placentals and marsupials) that has been determinant in the evolutionary explosion of modern mammals and their colonization of the world in an incredible variety of dietary niches.

ungulates. Herbivorous mammals with hoof-like nails. Ungulate mammals do not form a natural group. They include convergent, unrelated herbivorous clades, such as laurasiatherian euungulates (artiodactyls and perissodactyls) and afrotherian paenungulates (the ungulates of African origin: dassies, sea cows, and elephants).

zalambdodont teeth. The teeth of insectivorous mammals that are specialized puncturing teeth adapted for piercing insect shells. They are characterized by reduction of the tooth structures that have a crushing-grinding function (protofossa, talonid, and talonid basin) and of the metacone in upper molars.

Bibliography

Abbate, E., P. Bruni, M. P. Ferretti, C. Delmer, M. A. Laurenzi, M. Hagos, O. Bedri, L. Rook, M. Sagri, and Y. Libsekal. 2014. The East Africa Oligocene intertrappean beds: Regional distribution, depositional environments and Afro/Arabian mammal dispersals. Journal of African Earth Sciences 99:463–489.

Adaci, M., R. Tabuce, F. Mebrouk, M. Bensalah, P.-H. Fabre, L. Hautier, J.-J. Jaeger, V. Lazzari, M. Mahboubi, L. Marivaux, O. Otero, S. Peigné, and H. Tong. 2007. Nouveaux sites à vertébrés paléogènes dans la région des Gour Lazib (Sahara nord-occidental, Algérie). Comptes Rendus Palevol 6:535–544.

Adnet, S., H. Cappetta, and R. Tabuce. 2010. A middle–late Eocene vertebrate fauna (marine fish and mammals) from southwestern Morocco; preliminary report: Age and palaeobiogeographical implications. Geological Magazine 147:860–870.

Al-Kindi, M., M. Pickford, Y. Al-Sinani, I. Al-Ismaili, A. Hartman, and A. Heward. 2017. Large mammals from the Rupelian of Oman—recent finds. Fossil Imprint 73:300–321.

Álvarez-Carretero, S., A. U. Tamuri, M. Battini, F. F. Nascimento, E. Carlisle, R. J. Asher, Z. Yang, P. C. J. Donoghue, and M. dos Reis. 2022. A species-level timeline of mammal evolution integrating phylogenomic data. Nature 602:263–267.

Andrews, C. W. 1901a. Fossil vertebrates from Egypt. Zoologist 4:318–319.

Andrews, C. W. 1901b. Preliminary note on some recently discovered extinct vertebrates from Egypt (Part 1). Geological Magazine 8:400–409.

Andrews, C. W. 1901c. Über das Vorkommen von Proboscidiern in unter-tertiären Ablagerungen Aegyptens. Tageblatt Des V Internationalen Zoologischen Kongresses, Berlin 6:4–5.

Andrews, C. W. 1906. A Descriptive Catalogue of the Tertiary Vertebrata of the Fayûm, Egypt. Based on the Collection of the Egyptian Government in the Geological Museum, Cairo, and on the Collection in the British Museum (Natural History), London. British Museum, London. 324 pp.

Antoine, P.-O., L. Marivaux, D. A. Croft, G. Billet, M. Ganerød, C. Jaramillo, T. Martin, M. J. Orliac, J. Tejada, A. J. Altamirano, F. Duranthon, G. Fanjat, S. Rousse, and R. S. Gismondi. 2012. Middle Eocene rodents from Peruvian Amazonia reveal the pattern and timing of caviomorph origins and biogeography. Proceedings of the Royal Society B: Biological Sciences 279:1319–1326.

Archibald, J. D., and D. H. Deutschman. 2001. Quantitative analysis of the timing of the origin and diversification of extant placental orders. Journal of Mammalian Evolution 8:107–124.

Asher, R. 2019. Recent additions to the fossil record of tenrecs and golden moles. Afrotherian Conservation 1–13.

Barnosky, A. D., N. Matzke, S. Tomiya, G. O. U. Wogan, B. Swartz, T. B. Quental, C. Marshall, J. L. McGuire, E. L. Lindsey, K. C. Maguire, B. Mersey, and E. A. Ferrer. 2011. Has the Earth's sixth mass extinction already arrived? Nature 471:51–57.

Barrier, E., B. Vrielynck, F. Bergerat, M.-F. Brunet, J. Mosar, A. Poisson, and M. Sosson. 2008. Palaeotectonic maps of the Middle East: Tectono-sedimentary-palinspastic maps from late Norian to Pliocene. CGMW, Paris.

Barrow, E., E. R. Seiffert, and E. L. Simons. 2010. A primitive hyracoid (Mammalia, Paenungulata) from the early Priabonian (late Eocene) of Egypt. Journal of Systematic Palaeontology 8:213–244.

Barrow, E. C., E. R. Seiffert, and E. L. Simons. 2012. Cranial morphology of *Thyrohyrax domorictus* (Mammalia, Hyracoidea) from the early Oligocene of Egypt. Journal of Vertebrate Paleontology 32:166–179.

Beadnell, H. J. L. 1905. Topography and geology of the Fayum province of Egypt. Survey Dept., Egypt.

Beard, K. C. 1998. East of Eden: Asia as an important center of taxonomic origination in mammalian evolution. Bulletin of the Carnegie Museum of Natural History 34:5–39.

Beard, K. C. 2016. Out of Asia: Anthropoid origins and the colonization of Africa. Annual Review of Anthropology 45:199–213.

Beard, K. C., P. M. C. Coster, M. J. Salem, Y. Chaimanee, and J.-J. Jaeger. 2017. Biogeographic provincialism shown by Afro-Arabian mammals during the middle Cenozoic: Climate change, Red Sea rifting and global eustasy; pp. 48–68 in D. A. Agius, E. Khalil, E. Scerri, and A. Williams (eds.), Human Interaction with the Environment in the Red Sea. Brill, Leiden.

Benoit, J., S. Adnet, E. El Mabrouk, H. Khayati, M. Ben Haj Ali, L. Marivaux, G. Merzeraud, S. Merigeaud, M. Vianey-Liaud, and R. Tabuce. 2013. Cranial remain from Tunisia provides new clues for the origin and evolution of Sirenia (Mammalia, Afrotheria) in Africa. PLoS ONE 8:e54307.

Benoit, J., J.-Y. Crochet, M. Mahboubi, J.-J. Jaeger, M. Bensalah, M. Adaci, and R. Tabuce. 2016. New material of *Seggeurius amourensis* (Paenungulata, Hyracoidea), including a partial skull with intact basicranium. Journal of Vertebrate Paleontology 36:e1034358.

Bininda-Emonds, O. R., M. Cardillo, K. E. Jones, R. D. MacPhee, R. M. Beck, R. Grenyer, S. A. Price, R. A. Vos, J. L. Gittleman, and A. Purvis. 2007. The delayed rise of present-day mammals. Nature 446:507–512.

Blanckenhorn, M. 1903. Neue geologisch-stratigraphische Beobachtungen in Aegypten. Sitzungsberichte der Mathematisch-Physikalischen Classe der Königlichen Bayerischen Akademie der Wissenschaften, München, 32:353–433.

Boisserie, J.-R., F. Lihoreau, and M. Brunet. 2005. Origins of Hippopotamidae (Mammalia, Cetartiodactyla): Towards resolution. Zoologica Scripta 34:119–143.

Borths, M. R., and E. R. Seiffert. 2017. Craniodental and humeral morphology of a new species of *Masrasector* (Teratodontinae, Hyaenodonta, Placentalia) from the late Eocene of Egypt and locomotor diversity in hyaenodonts. PLoS ONE 12:e0173527.

Bown, T., and E. L. Simons. 1984. First record of marsupial (Metatheria: Polyprotodonta) from Oligocene in Africa. Nature 308:447–449.

Bown, T. M., and E. L. Simons. 1987. New Oligocene Ptolemaiidae (Mammallia:?Pantolesta) from the Jebel Qatrani Formation, Fayum Depression, Egypt. Journal of Vertebrate Paleontology 7:311–324.

Bronner, G. N., S. Mynhardt, N. C. Bennett, L. Cohen, N. Crumpton, M. Hofreiter, P. Arnold, and R. J. Asher. 2024. Phylogenetic history of golden moles and tenrecs (Mammalia: Afrotheria). Zoological Journal of the Linnean Society 201:184–213.

Cappetta, H., J.-J. Jaeger, M. Sabatier, B. Sige, J. Sudre, and M. Vianey-Liaud. 1978. Decouverte dans le Paléocène du Maroc des plus anciens mammiferes eutheriens d'Afrique. Geobios 11:257–263.

Clemens, W. A. 2010. Were immigrants a significant part of the earliest Paleocene mammalian fauna of the North American Western Interior? Vertebrata Pal-Asiatica 48:285–307.

Clementz, M. T., P. A. Holroyd, and P. L. Koch. 2008. Identifying aquatic habits of herbivorous mammals through stable isotope analysis. PALAIOS 23:574–585.

Coiffait, P.-E., B. Coiffait, J.-J. Jaeger, and M. Mahboubi. 1984. Un nouveau gisement d'âge Eocène supérieur sur le versant sud des Nementcha (Algérie orientale): découverte des plus anciens rongeurs d'Afrique. Comptes Rendus de l'Académie Des Sciences de Paris 299, II:893–898.

Coryndon, S., and R. J. G. Savage. 1973. The origin and affinities of African mammal faunas. Special Papers in Palaeontology 12:121–135.

Coster, P., M. Benammi, M. Salem, A. A. Bilal, Y. Chaimanee, X. Valentin, M. Brunet, and J.-J. Jaeger. 2012. New hystricognathous rodents from the early Oligocene of central Libya (Zallah Oasis, Sahara Desert): Systematic, phylogenetic, and biochronologic implications. Annals of Carnegie Museum 80:239–259.

Coster, P. M. C., K. C. Beard, M. J. Salem, Y. Chaimanee, M. Brunet, and J.-J. Jaeger. 2015. A new early Oligocene mammal fauna from the Sirt Basin, central Libya: Biostratigraphic and paleobiogeographic implications. Journal of African Earth Sciences 104:43–55.

Court, N. 1992. A unique form of dental bilophodonty and a functional interpretation of peculiarities in the masticatory system of *Arsinoitherium* (Mammalia, Embrithopoda). Historical Biology 6:91–111.

Court, N. 1994. Limb posture and gait in *Numidotherium koholense*, a primitive proboscidean from the Eocene of Algeria. Zoological Journal of the Linnean Society 111:297–338.

Court, N., and M. Mahboubi. 1993. Reassessment of lower Eocene *Seggeurius amourensis*: Aspects of primitive dental morphology in the mammalian order Hyracoidea. Journal of Paleontology 67:889–893.

Couvreur, T. L. P., G. Dauby, A. Blach-Overgaard, V. Deblauwe, S. Dessein, V. Droissart, O. J. Hardy, D. J. Harris, S. B. Janssens, A. C. Ley, B. A. Mackinder, B. Sonké, M. S. M. Sosef, T. Stévart, J.-C. Svenning, J. J. Wieringa, A. Faye, A. D. Missoup, K. A. Tolley, V. Nicolas, S. Ntie, F. Fluteau, C. Robin, F. Guillocheau, D. Barboni, and P. Sepulchre. 2021. Tectonics, climate and the diversification of the tropical African terrestrial flora and fauna. Biological Reviews 96:16–51.

Crespo, V. D., F. J. Goin, and M. Pickford. 2022. The last African metatherian. Fossil Record 25:173–186.

Dames, W. B. 1883. Über eine tertiäre Wirbelthierfauna von der westlichen Insel des Birket-el-Qurun im Fajum (Aegypten). Sitzungsberichte Der Königlich Preussischen Akademie Der Wissenschaften Zu Berlin 1:1229–153.

Dames, W. B. 1894. Über Zeuglodonten aus Aegypten und die Beziehungen der Archaeoceten zu den übrigen Cetaceen. Geologische Und Palaontologische Abhandlungen, Jena, 5:189–222.

Dartevelle, E. 1935. Les premiers restes de mammifères du Tertiaire du Congo: la faune Miocene de Malembe. 2nd Cong. Nat. Sc. Bruxelles, CR 1:715–720.

De Blieux, D. D., M. R. Baumrind, E. L. Simons, P. S. Chatrath, G. E. Meyer, and Y. S. Attia. 2006. Sexual dimorphism of the internal mandibular chamber in Fayum Pliohyracidae (Mammalia). Journal of Vertebrate Paleontology 26:160–169.

De Blieux, D. D., and E. L. Simons. 2002. Cranial and dental anatomy of *Antilohyrax pectidens:* A late Eocene hyracoid (Mammalia) from the Fayum, Egypt. Journal of Vertebrate Paleontology 22:122–136.

De Bonis, L., J.-J. Jaeger, B. Coiffait, and P.-E. Coiffait. 1988. Découverte du plus ancien primate catarrhinien connu dans l'Eocene supérieur d'Afrique du Nord. CR Acad. Sci. Paris 306:929–934.

de Vries, D., S. Heritage, M. R. Borths, H. M. Sallam, and E. R. Seiffert. 2021. Widespread loss of mammalian lineage and dietary diversity in the early Oligocene of Afro-Arabia. Communications Biology 4:1172.

Delmer, C. 2005. Les première phases de differenciation des proboscidiens (Tethytheria, Mammalia): le rôle du *Barytherium grave* de Libye. Unpublished PhD thesis, Muséum National d'Histoire Naturelle de Paris, Paris, 470 pp.

Delmer, C. 2009. Reassessment of the generic attribution of *Numidotherium savagei* and the homologies of lower incisors in proboscideans. Acta Palaeontologica Polonica 54:561–580.

Dercourt, J., L.-E. Ricou, and B. Vrielynck. 1993. Atlas Tethys Palaeoenvironmental Maps, CCGM. Gauthier-Villars, Paris.

Domning, D. P., P. D. Gingerich, E. L. Simons, and F. A. Ankel-Simons. 1994. A new early Oligocene dugongid (Mammalia, Sirenia) from Fayum Province, Egypt. Contributions from the Museum of Paleontology, the University of Michigan, 29:89–108.

Ducrocq, S., J.-R. Boisserie, J.-J. Tiercelin, C. Delmer, G. Garcia, M. F. Kyalo, M. G. Leakey, L. Marivaux, O. Otero, S. Peigné, P. Tassy, and F. Lihoreau. 2010. New Oligocene vertebrate localities from Northern Kenya (Turkana basin). Journal of Vertebrate Paleontology 30:293–299.

Fabre, P.-H., M.-K. Tilak, C. Denys, P. Gaubert, V. Nicolas, E. J. P. Douzery, and L. Marivaux. 2018. Flightless scaly-tailed squirrels never learned how to fly: A reappraisal of Anomaluridae phylogeny. Zoologica Scripta 47:404–417.

Filhol, H. 1878. Note sur la découverte d'un nouveau Mammifère marin (*Manatus coulombi*) en Afrique dans les carrières de Mokattam près du Caire. Bulletin de La Société Philomathique de Paris 7:124–125.

Fleagle, J. 2013. Primate Adaptation and Evolution. Academic Press, Amsterdam.

Gagnon, M. 1997. Ecological diversity and community ecology in the Fayum sequence (Egypt). Journal of Human Evolution 32:133–160.

Gaudry, A. 1891. Quelques remarques sur les mastodontes à propos de l'animal du Cherichira. Mémoire de la Société Géologique de France 8, II:1–6.

Gebo, D. L., and D. T. Rasmussen. 1985. The earliest fossil pangolin (Pholidota: Manidae) from Africa. Journal of Mammalogy 66:538–541.

Gentry, A., and J. Hooker. 1988. The phylogeny of the Artiodactyla; pp. 235–272 in M. Benton (ed.), The Phylogeny and Classification of the Tetrapods. Vol. 2: Mammals. Systematics Association Special Volume no. 35B. Clarendon Press, Oxford.

Gevin, P., R. Lavocat, N. Mongereau, and J. Sudre. 1975. Découverte de Mammifères dans la moitié inférieure de l'Eocène continental du Nord Ouest du Sahara. Comptes Rendus de l'Académie Des Sciences de Paris II:244–249.

Gheerbrant, E. 1987. Les vertébrés continentaux de l'Adrar Mgorn (Maroc, Paléocène); une dispersion de mammifères transtéthysienne aux environs de la limite Mesozoïque/Cénozoïque? Geodinamica Acta 1:233–245.

Gheerbrant, E. 1989. Les mammifères paléocènes du bassin d'Ouarzazate (Maroc): étude systématique, phylogénique, paléoécologique et paléobiogéographique des plus anciens placentaires d'Afrique. Université P. & M. Curie, Paris, 473 pp.

Gheerbrant, E. 1992. Les mammifères paléocènes du Bassin d'Ouarzazate (Maroc). I. Introduction générale et Palaeoryctidae. Palaeontographica, A 224:67–132.

Gheerbrant, E. 1994. Les mammifères paléocènes du Bassin d'Ouarzazate (Maroc). II. Todralestidae (Eutheria, Proteutheria). Palaeontographica, A 231:133–188.

Gheerbrant, E. 1995. Les mammifères paléocènes du Bassin d'Ouarzazate (Maroc). III. Adapisoriculidae et autres mammifères (Carnivora, ?Creodonta, Condylarthra, ?Ungulata et *incertae sedis*). Palaeontographica, A 237:39–132.

Gheerbrant, E. 2009. Paleocene emergence of elephant relatives and the rapid radiation of African ungulates. Proceedings of the National Academy of Sciences 106:10717–10721.

Gheerbrant, E., M. Amaghzaz, B. Bouya, F. Goussard, and C. Letenneur. 2014. *Ocepeia* (middle Paleocene of Morocco): The oldest skull of an Afrotherian mammal. PLoS ONE 9:e89739.

Gheerbrant, E., G. Billet, and M. Pickford. 2025. New data on the earliest known arsinoitheriid embrithopod (Mammalia, Paenungulata), *Namatherium* from the middle Eocene of Namibia. Geodiversitas, in press.

Gheerbrant, E., B. Bouya, and M. Amaghzaz. 2012. Dental and cranial anatomy of *Eritherium azzouzorum* from the Paleocene of Morocco, earliest known proboscidean mammal. Palaeontographica, A 297:151–183.

Gheerbrant, E., A. Filippo, and A. Schmitt. 2016. Convergence of afrotherian and laurasiatherian ungulate-like mammals: First morphological evidence from the Paleocene of Morocco. PLoS One 11:1–35.

Gheerbrant, E., F. Khaldoune, A. Schmitt, and R. Tabuce. 2021. Earliest embrithopod mammals (Afrotheria, Tethytheria) from the early Eocene of Morocco: Anatomy, systematics and phylogenetic significance. Journal of Mammal Evolution 28:245–283.

Gheerbrant, E., S. Peigné, and H. Thomas. 2007. Première description du squelette d'un Mammifère Hyracoïde du Paléogène: *Saghatherium antiquum* de l'Oligocène inférieur de Jebel al Hasawnah, Libye. Palaeontographica Abteilung A 279:93–145.

Gheerbrant, E., and J.-C. Rage. 2006. Paleobiogeography of Africa: How distinct from Gondwana and Laurasia? Palaeogeography, Palaeoclimatology, Palaeoecology 241:224–246.

Gheerbrant, E., A. Schmitt, and L. Kocsis. 2018. Early African fossils elucidate the origin of embrithopod mammals. Current Biology 28:2167–2173.

Gheerbrant, E., J. Sudre, and H. Cappetta. 1996. A Palaeocene proboscidean from Morocco. Nature 383:68–70.

Gheerbrant, E., J. Sudre, Cappetta, H., and G. Bignot. 1998. *Phosphatherium escuilliei* du Thanétien du bassin des Ouled Abdoun (Maroc), plus ancien proboscidien (Mammalia) d'Afrique. Geobios 30:247–269.

Gheerbrant, E., J. Sudre, H. Cappetta, M. Iarochéne, M. Amaghzaz, and B. Bouya. 2002. A new large mammal from the Ypresian of Morocco: Evidence of surprising diversity of early proboscideans. Acta Palaeontologica Polonica 47:493–506.

Gheerbrant, E., J. Sudre, M. Iarochene, and A. Moumni. 2001. First ascertained African "Condylarth" mammals (primitive ungulates: cf. Bulbulodentata and cf. Phenacodonta) from the earliest Ypresian of the Ouled Abdoun Basin, Morocco. Journal of Vertebrate Paleontology 21:107–118.

Gheerbrant, E., J. Sudre, P. Tassy, M. Amaghzaz, B. Bouya, and M. Iarochene. 2005. Nouvelles données sur *Phosphatherium escuilliei* (Mammalia, Proboscidea) de l'Eocène inférieur du Maroc, apports à la phylogénie des Proboscidea et des ongulés lophodontes. Geodiversitas 27:239–333.

Gheerbrant, E., and P. Tassy. 2009. L'origine et l'évolution des éléphants. Comptes Rendus Palevol 8:281–294.

Gheerbrant, E., H. Thomas, J. Roger, S. Sen, and Z. Al Sulaimani. 1993. Deux nouveaux primates dans l'Oligocène inférieur de Taqah (Sultanat d'Oman): premiers adapiformes (?Anchomomyini) de la Péninsule Arabique? Palaeovertebrata 22:141–196.

Gheerbrant, E., H. Thomas, S. Sen, and Z. Al-Sulaimani. 1995. Nouveau Primate Oligopithecinae (Simiiformes) de l'Oligocène inférieur de Taqah, Sultanat d'Oman. Comptes Rendus de l'Académie Des Sciences de Paris II a:425–432.

Gingerich, P. D. 2006. Environment and evolution through the Paleocene–Eocene Thermal Maximum. Trends in Ecology & Evolution 21:246–253.

Gingerich, P. D. 2024. Wadi Al-Hitan or "Valley of Whales"—an Eocene World Heritage Site in the Western Desert of Egypt. Geological Society, London, Special Publications 543:SP543-2022-203.

Godinot, M. 2006. Lemuriform origins as viewed from the fossil record. Folia Primatologica 77:446–464.

Godinot, M., and M. Mahboubi. 1994. Les petits primates simiiformes de Glib Zegdou (Eocène inférieur à moyen d'Algérie). CR Acad. Sci. Paris 319:357–364.

Godinot, M., B. Senut, and M. Pickford. 2018. Primitive Adapidae from Namibia sheds light on the early primate radiation in Africa. Communications of the Geological Survey of Namibia 18:140–162.

Gorodiski, A., and R. Lavocat. 1953. Premiere decouverte de mammiferes dans le Tertiaire (Lutetien) du Senegal. Comptes Rendus Sommaires de La Société Géologique de France 15:314–316.

Goswami, A., G. V. R. Prasad, P. Upchurch, D. Boyer, E. Seiffert, O. Verma, E. Gheerbrant, and J. J. Flynn. 2011. A radiation of arboreal basal eutherian mammals beginning in the Late Cretaceous of India. Proceeding of the National Academy of Sciences 108:16333–16338.

Grohé, C., M. Morlo, Y. Chaimanee, C. Blondel, P. Coster, X. Valentin, M. Salem, A. A. Bilal, J.-J. Jaeger, and M. Brunet. 2012. New Apterodontinae (Hyaenodontida) from the Eocene locality of Dur At-Talah (Libya): Systematic, paleoecological and phylogenetical implications. PLoS ONE 7:e49054.

Gunnell, G. F., D. M. Boyer, A. R. Friscia, S. Heritage, F. K. Manthi, E. R. Miller, H. M. Sallam, N. B. Simmons, N. J. Stevens, and E. R. Seiffert. 2018. Fossil lemurs from Egypt and Kenya suggest an African origin for Madagascar's aye-aye. Nature Communications 9:1–12.

Gunnell, G. F., P. D. Gingerich, and P. A. Holroyd. 2010. Ptolemaiida; pp. 83–87 in L. Werdelin and W. J. Sanders (eds.), Cenozoic Mammals of Africa. University of California Press, Berkeley.

Gunnell, G. F., B. F. Jacobs, P. S. Herendeen, J. J. Head, E. Kowalski, C. P. Msuya, F. A. Mizambwa, T. Harrison, J. Habersetzer, and G. Storch. 2003. Oldest placental mammal from sub-Saharan Africa: Eocene microbat from Tanzania—evidence for early evolution of sophisticated echolocation. Palaeontologia Electronica 5:1–10.

Gunnell, G. F., N. B. Simmons, and E. R. Seiffert. 2014. New Myzopodidae (Chiroptera) from the late Paleogene of Egypt: Emended family diagnosis and biogeographic origins of Noctilionoidea. PLoS ONE 9:e86712.

Haddoumi, H., R. Allain, S. Meslouh, G. Metais, M. Monbaron, D. Pons, J.-C. Rage, R. Vullo, S. Zouhri, and E. Gheerbrant. 2016. Guelb el Ahmar (Bathonian, Anoual Syncline, eastern Morocco): First continental flora and fauna including mammals from the Middle Jurassic of Africa. Gondwana Research 29:290–319.

Halliday, T. J. D., P. Upchurch, and A. Goswami. 2017. Resolving the relationships of Paleocene placental mammals: Paleocene mammal phylogeny. Biological Reviews 92:521–550.

Hartenberger, J.-L. 1986. Hypothèse paléontologique sur l'origine des Macroscelidea (Mammalia). CR Acad. Sci. Paris, Ser. II 302:247–249.

Hartenberger, J.-L., and B. Marandat. 1992. A new genus and species of an early Eocene primate from North Africa. Human Evolution 7:9–16.

Hartenberger, J.-L., C. Martinez, and A. B. Said. 1985. Découverte de mammifères d'âge Eocène inférieur en Tunisie centrale. Comptes Rendus de l'Académie Des Sciences. Série 2, Mécanique, Physique, Chimie, Sciences de l'univers, Sciences de La Terre 301:649–652.

Harzhauser, M., A. Kroh, O. Mandic, W. E. Piller, U. Göhlich, M. Reuter, and B. Berning. 2007. Biogeographic responses to geodynamics: A key study all around the Oligo–Miocene Tethyan Seaway. Zoologischer Anzeiger: A Journal of Comparative Zoology 246:241–256.

Hautier, L., R. Sarr, R. Tabuce, F. Lihoreau, S. Adnet, D. P. Domning, M. Samb, and P. M. Hameh. 2012. First prorastomid sirenian from Senegal (Western Africa) and the Old World origin of sea cows. Journal of Vertebrate Paleontology 32:1218–1222.

Hautier, L., R. Tabuce, M. J. Mourlam, K. E. Kassegne, Y. Z. Amoudji, M. Orliac, F. Quillévéré, A.-L. Charruault, A. K. C. Johnson, and G. Guinot. 2021. New middle Eocene proboscidean from Togo illuminates the early evolution of the elephantiform-like dental pattern. Proceedings of the Royal Society B: Biological Sciences 288:20211439.

Heritage, S., D. Fernández, H. M. Sallam, D. T. Cronin, J. M. E. Echube, and E. R. Seiffert. 2016. Ancient phylogenetic divergence of the enigmatic African rodent Zenkerella and the origin of anomalurid gliding. PeerJ 4:e2320.

Heritage, S., E. R. Seiffert, and M. R. Borths. 2021. Recommended fossil calibrators for time-scaled molecular phylogenies of Afrotheria. Afrotherian Conservation 9–13.

Hooker, J. J., M. R. Sánchez-Villagra, F. J. Goin, E. L. Simons, Y. Attia, and E. R. Seiffert. 2008. The origin of Afro-Arabian "didelphimorph" marsupials. Palaeontology 51:635–648.

Huber, B. T., K. G. MacLeod, D. K. Watkins, and M. F. Coffin. 2018. The rise and fall of the Cretaceous Hot Greenhouse climate. Global and Planetary Change 167:1–23.

Jacobs, B. F., and P. S. Herendeen. 2004. Eocene dry climate and woodland vegetation in tropical Africa reconstructed from fossil leaves from northern Tanzania. Palaeogeography, Palaeoclimatology, Palaeoecology 213:115–123.

Jaeger, J.-J., K. C. Beard, Y. Chaimanee, M. Salem, M. Benammi, O. Hlal, P. Coster, A. A. Bilal, P. Duringer, M. Schuster, X. Valentin, B. Marandat, L. Marivaux, E. Métais, O. Hammuda, and M. Brunet. 2010. Late middle Eocene epoch of Libya yields earliest known radiation of African anthropoids. Nature 467:1095–1098.

Jaeger, J.-J., C. Denys, and B. Coiffait. 1985. New Phiomorpha and Anomaluridae from the late Eocene of North-West Africa: Phylogenetic implications; pp. 567–588 in Evolutionary Relationships Among Rodents. Springer, New York.

Jaeger, J.-J., L. Marivaux, M. Salem, A. A. Bilal, M. Benammi, Y. Chaimanee, P. Duringer, B. Marandat, E. Métais, M. Schuster, X. Valentin, and M. Brunet. 2010. New rodent assemblages from the Eocene Dur At-Talah escarpment (Sahara of central Libya): Systematic, biochronological, and palaeobiogeographical implications. Zoological Journal of the Linnean Society 160:195–213.

Jaeger, J.-J., M. Salem, A. Abolhassan Bilal, M. Benammi, Y. Chaimanee, P. Duringer, B. Marandat, L. Marivaux, E. Métais, M. Schuster, and M. Brunet. 2012. Dur at Talah vertebrate locality revisited: New data on its stratigraphy, age, sedimentology and mammalian fossils. Geology of Southern Libya 1:241–258.

Kampouridis, P., J. Hartung, F. J. Augustin, H. El Atfy, and G. S. Ferreira. 2023. Dental eruption and adult dentition of the enigmatic ptolemaiid *Qarunavus meyeri* from the Oligocene of the Fayum Depression (Egypt) revealed by micro-computed tomography clarifies its phylogenetic position. Zoological Journal of the Linnean Society 199:1078–1091.

Kappelman, J., D. Tab Rasmussen, W. J. Sanders, M. Feseha, T. Bown, P. Copeland, J. Crabaugh, J. Fleagle, M. Glantz, A. Gordon, B. Jacobs, M. Maga, K. Muldoon, A. Pan, L. Pyne, B. Richmond, T. Ryan, E. R. Seiffert, S. Sen, L. Todd, M. C. Wiemann, and A. Winkler. 2003. Oligocene mammals from Ethiopia and faunal exchange between Afro-Arabia and Eurasia. Nature 426:549–552.

Kielan-Jaworowska, Z., R. L. Cifelli, and Z.-X. Luo. 2004. Mammals from the Age of Dinosaurs: Origins, Evolution, and Structure. Columbia University Press, New York, 656 pp.

Kocsis, Á. T., and C. R. Scotese. 2021. Mapping paleocoastlines and continental flooding during the Phanerozoic. Earth-Science Reviews 213:103463.

Kron, D. D., and E. Manning. 1998. Anthracotheriidae; pp. 381–389 in C. Janis (ed.), Evolution of Tertiary Mammals of North America. Vol. 1: Terrestrial Carnivores, Ungulates, and Ungulatelike Mammals. Cambridge University Press, Cambridge.

Ksepka, D. T., T. A. Stidham, and T. E. Williamson. 2017. Early Paleocene landbird supports rapid phylogenetic and morphological diversification of crown birds after the K–Pg mass extinction. Proceedings of the National Academy of Sciences 114:8047–8052.

Kumar, S., and S. B. Hedges. 1998. A molecular timescale for vertebrate evolution. Nature 392:917–920.

Lasseron, M. 2020. Paléobiodiversité, évolution et paléobiogéographie des vertébrés mésozoïques africains et gondwaniens: apport des gisements du Maroc oriental. PhD thesis, Sorbonne Université–MNHN, Paris, 542 pp.

Lasseron, M., R. Allain, E. Gheerbrant, H. Haddoumi, N.-E. Jalil, G. Métais, J.-C. Rage, R. Vullo, and S. Zouhri. 2019. New data on the microvertebrate fauna from the Upper Jurassic or lowest Cretaceous of Ksar Metlili (Anoual Syncline, eastern Morocco). Geological Magazine 157:367–392.

Leduc, P. 1996. Caractéristiques évolutives de faunes de mammifères d'Europe Occidentale et d'Amérique du Nord au Paléogène. Doctorat de l'Université Paris 6, Université Pierre & Marie Curie, Paris, 453 pp.

Licht, A., G. Métais, P. Coster, D. İbilioğlu, F. Ocakoğlu, J. Westerweel, M. Mueller, C. Campbell, S. Mattingly, M. C. Wood, and K. C. Beard. 2022. Balkanatolia: The insular mammalian biogeographic province that partly paved the way to the Grande Coupure. Earth-Science Reviews 226:103929.

Lihoreau, F., J.-R. Boisserie, F. K. Manthi, and S. Ducrocq. 2015. Hippos stem from the longest sequence of terrestrial cetartiodactyl evolution in Africa. Nature Communications 6:6264.

Liu, A. G. S. C., E. R. Seiffert, and E. L. Simons. 2008. Stable isotope evidence for an amphibious phase in early proboscidean evolution. Proceedings of the National Academy of Sciences 105:5786–5791.

López-Torres, S., K. R. Selig, A. M. Burrows, and M. T. Silcox. 2020. The toothcomb of *Karanisia clarki*: Was this species an exudate-feeder?; pp. 67–75 in K. A. I. Nekaris and A. M. Burrows (eds.), Evolution, Ecology and Conservation of Lorises and Pottos, 1st ed. Cambridge University Press.

Madsen, O., M. Scally, C. J. Douady, D. J. Kao, R. W. DeBry, R. Adkins, H. M. Amrine, M. J. Stanhope, W. W. de Jong, and M. S. Springer. 2001. Parallel adaptive radiations in two major clades of placental mammals. Nature 409:610–614.

Mahboubi, M., R. Ameur, J.-Y. Crochet, and J. J. Jaeger. 1984. Earliest known proboscidean from early Eocene of north-west Africa. Nature 308:543–544.

Mahboubi, M., R. Ameur, J.-Y. Crochet, and J. J. Jaeger. 1986. El Kohol (Saharan Atlas, Algeria): A new Eocene mammal locality in northwestern Africa. Stratigraphical, phylogenetic and paleobiogeographical data. Palaeontographica. Abteilung A, Paläozoologie, Stratigraphie 192:15–49.

Marivaux, L., M. Adaci, M. Bensalah, H. G. Rodrigues, L. Hautier, M. Mahboubi, F. Mebrouk, R. Tabuce, and M. Vianey-Liaud. 2011. Zegdoumyidae (Rodentia, Mammalia), stem anomaluroid rodents from the early to middle Eocene of Algeria (Gour Lazib, Western Sahara): New dental evidence. Journal of Systematic Palaeontology 9:563–588.

Marivaux, L., S. Adnet, M. Benammi, R. Tabuce, and M. Benammi. 2017. Anomaluroid rodents from the earliest Oligocene of Dakhla, Morocco, reveal the long-lived and morphologically conservative pattern of the Anomaluridae and

Nonanomaluridae during the Tertiary in Africa. Journal of Systematic Palaeontology 15:539–569.

Marivaux, L., S. Adnet, M. Benammi, R. Tabuce, J. Yans, and M. Benammi. 2017. Earliest Oligocene hystricognathous rodents from the Atlantic margin of northwestern Saharan Africa (Dakhla, Morocco): Systematic, paleobiogeographical, and paleoenvironmental implications. Journal of Vertebrate Paleontology 37:e1357567.

Marivaux, L., E. M. Essid, W. Marzougui, H. Khayati Ammar, S. Adnet, B. Marandat, G. Merzeraud, R. Tabuce, and M. Vianey-Liaud. 2014. A new and primitive species of *Protophiomys* (Rodentia, Hystricognathi) from the late middle Eocene of Djebel el Kébar, Central Tunisia. Palaeovertebrata 38: 12–22.

Marivaux, L., E. M. Essid, W. Marzougui, H. Khayati Ammar, S. Adnet, B. Marandat, G. Merzeraud, A. Ramdarshan, R. Tabuce, and M. Vianey-Liaud. 2014. A morphological intermediate between eosimiiform and simiiform primates from the late middle Eocene of Tunisia: Macroevolutionary and paleobiogeographic implications of early anthropoids. American Journal of Physical Anthropology 154:387–401.

Marivaux, L., A. Ramdarshan, E. M. Essid, W. Marzougui, H. K. Ammar, R. Lebrun, B. Marandat, G. Merzeraud, R. Tabuce, and M. Vianey-Liaud. 2013. *Djebelemur*, a tiny pre-tooth-combed primate from the Eocene of Tunisia: A glimpse into the origin of crown strepsirhines. PLoS ONE 8:e80778.

Mason, M. J., N. C. Bennett, and M. Pickford. 2018. The middle and inner ears of the Palaeogene golden mole Namachloris: A comparison with extant species. Journal of Morphology 279:375–395.

Mattingly, S. G., K. C. Beard, P. M. C. Coster, M. J. Salem, Y. Chaimanee, and J.-J. Jaeger. 2020. A new carnivoraform from the early Oligocene of Libya: Oldest known record of Carnivoramorpha in Africa. Journal of African Earth Sciences 172:103994.

McKenna, M. C. 1975. Toward a phylogenetic classification of the Mammalia; pp. 21–46 in W. P. Luckett and F. S. Szalay (eds.), Phylogeny of the Primates: A Multidisciplinary Approach. Springer US, Boston.

Métais, G., P. Coster, M. Kaya, A. Licht, K. Miller, F. Ocakoğlu, K. Rust, and K. C. Beard. 2024. Rapid colonization and diversification of a large-bodied mammalian herbivore clade in an insular context: New embrithopods from the Eocene of Balkanatolia. Journal of Mammalian Evolution 31:15.

Montgelard, C., E. Forty, V. Arnal, and C. A. Matthee. 2008. Suprafamilial relationships among Rodentia and the phylogenetic effect of removing fast-evolving nucleotides in mitochondrial, exon and intron fragments. BMC Evolutionary Biology 8:1–16.

Morley, R. J. 2000. Origin and Evolution of Tropical Rain Forests. John Wiley & Sons, Chichester, UK, pp. 362.

Murphy, W. J., E. Eizirik, W. E. Johnson, Y. P. Zhang, O. A. Ryder, and S. J. O'Brien. 2001. Molecular phylogenetics and the origins of placental mammals. Nature 409:614–618.

Noubhani, A., L. Hautier, J.-J. Jaeger, M. Mahboubi, and R. Tabuce. 2008. Variabilité dentaire et crânienne de *Numidotherium koholense* (Mammalia, Proboscidea) de l'Éocène d'El Kohol, Algérie. Geobios 41:515–531.

O'Leary, M. A., J. I. Bloch, J. J. Flynn, T. J. Gaudin, A. Giallombardo, N. P. Giannini, S. L. Goldberg, B. P. Kraatz, Z.-X. Luo, J. Meng, X. Ni, M. J. Novacek, F. A. Perini, Z. S. Randall, G. W. Rougier, E. J. Sargis, M. T. Silcox, N. B. Simmons, M. Spaulding, P. M. Velazco, M. Weksler, J. R. Wible, and A. L. Cirranello. 2013. The placental mammal ancestor and the post-K–Pg radiation of placentals. Science 339:662–667.

Olson, E. R., M. R. Carlson, V. M. S. Ramanujam, L. Sears, S. E. Anthony, P. S. Anich, L. Ramon, A. Hulstrand, M. Jurewicz, A. S. Gunnelson, A. M. Kohler, and J. G. Martin. 2021. Vivid biofluorescence discovered in the nocturnal Springhare (Pedetidae). Scientific Reports 11:4125.

Orliac, M., J.-R. Boisserie, L. MacLatchy, and F. Lihoreau. 2010. Early Miocene hippopotamids (Cetartiodactyla) constrain the phylogenetic and spatiotemporal settings of hippopotamid origin. Proceedings of the National Academy of Sciences 107:11871–11876.

Osborn, H. F. 1908. New fossil mammals from the Fayûm Oligocene, Egypt. Bulletin of the American Museum of Natural History 24:265–272.

Osborn, H. F. 1909. New carnivorous mammals from the Fayûm Oligocene, Egypt. Bulletin of the American Museum of Natural History 26:415–424.

Owen, R. 1875. On fossil evidences of a sirenian mammal (*Eotherium aegyptiacum*, Owen) from the Nummulitic Eocene of the Mokattam Cliffs, near Cairo. Quarterly Journal of the Geological Society 31:100–105.

Patterson, B. D., and N. S. Upham. 2014. A newly recognized family from the Horn of Africa, the Heterocephalidae (Rodentia: Ctenohystrica). Zoological Journal of the Linnean Society 172:942–963.

Phillips, M. J. 2016. Geomolecular dating and the origin of placental mammals. Systematic Biology 65:546–557.

Phillips, M. J., and C. Fruciano. 2018. The soft explosive model of placental mammal evolution. BMC Evolutionary Biology 18:1–13.

Pickford, M. 1986. Première découverte d'une faune mammalienne terrestre paléogène d'Afrique sub-saharienne. Comptes Rendus de l'Académie Des Sciences. Série 2, Mécanique, Physique, Chimie, Sciences de l'univers, Sciences de La Terre 302:1205–1210.

Pickford, M. 1987. Recognition of an early Oligocene or late Eocene mammal fauna from Cabinda Angola. Mus. Roy. Afr. Centr., Tervuren (Belg.), Dépt. Géol. Min., Rapp. Ann. 1985–1986:89–92.

Pickford, M. 2008. The myth of the hippo-like anthracothere: The eternal problem of homology and convergence. Spanish Journal of Palaeontology 23:31–90.

Pickford, M. 2015a. *Bothriogenys* (Anthracotheriidae) from the Bartonian of Eoridge, Namibia. Communications of the Geological Survey of Namibia 16:215–222.

Pickford, M. 2015b. Cenozoic geology of the Northern Sperrgebiet, Namibia, accenting the Palaeogene. Communications of the Geological Survey of Namibia 16:10–104.

Pickford, M. 2015c. Chrysochloridae (Mammalia) from the Lutetian (middle Eocene) of Black Crow, Namibia. Communications of the Geological Survey of Namibia 16:112–120.

Pickford, M. 2015d. Late Eocene Chrysochloridae (Mammalia) from the Sperrgebiet, Namibia. Communications of the Geological Survey of Namibia 16:160–199.

Pickford, M. 2015e. New Titanohyracidae (Hyracoidea: Afrotheria) from the late Eocene of Namibia. Communications of the Geological Survey of Namibia 16:200–214.

Pickford, M. 2018a. *Diamantochloris* mandible from the Ypresian/Lutetian of Namibia. Communications of the Geological Survey of Namibia 19:51–65.

Pickford, M. 2018b. Fossil fruit bat from the Ypresian/Lutetian of Black Crow, Namibia. Communications of the Geological Survey of Namibia 18:64–71.

Pickford, M. 2019. New Chrysochloridae (Mammalia) from the middle Eocene of Black Crow, Namibia. Communications of the Geological Survey of Namibia 21:40–47.

Pickford, M., B. Senut, J. Morales, P. Mein, and I. Sanchez. 2008. Mammalia from the Lutetian of Namibia. Memoirs of the Geological Survey of Namibia 20:465–514.

Poblete, F., G. Dupont-Nivet, A. Licht, D. J. Van Hinsbergen, P. Roperch, M. Mihaly-nuk, S. Johnston, F. Guillocheau, G. Baby, and F. Fluteau. 2021. Towards interactive global paleogeographic maps, new reconstructions at 60, 40 and 20 Ma. Earth-Science Reviews 214:103508.

Prasad, G. V. R., J. J. Jaeger, A. Sahni, E. Gheerbrant, and C. K. Khajuria. 1994. Eutherian mammals from the Upper Cretaceous (Maastrichtian) Intertrappean Beds of Naskal, Andhra Pradesh, India. Journal of Vertebrate Paleontology 14:260–277.

Rage, J.-C. 1996. Le peuplement animal de Madagascar: une composante venue de Laurasie est-elle envisageable?; pp. 27–35 in W. R. Lourenço (ed.), Biogéographie de Madagascar. Publications de l'IRD.

Rage, J.-C., and E. Gheerbrant. 2020. Island Africa and vertebrate evolution: A review of data and working hypotheses; pp. 251–264 in G. V. R. Prasad and R. Patnaik (eds.), Biological Consequences of Plate Tectonics: New Perspectives on Post-Gondwana Break-Up—a Tribute to Ashok Sahni. Vertebrate Paleobiology and Paleoanthropology Series. Springer International Publishing, Cham.

Rasmussen, D. T., T. M. Bown, and E. L. Simons. 1992. The Eocene–Oligocene transition in continental Africa; pp. 548–566 in Eocene-Oligocene Climatic and Biotic Evolution. Princeton University Press.

Rasmussen, D. T., G. C. Conroy, and E. L. Simons. 1998. Tarsier-like locomotor specializations in the Oligocene primate Afrotarsius. Proceedings of the National Academy of Sciences 95:14848–14850.

Rasmussen, D. T., A. R. Friscia, M. Gutierrez, J. Kappelman, E. R. Miller, S. Muteti, D. Reynoso, J. B. Rossie, T. L. Spell, N. J. Tabor, E. Gierlowski-Kordesch, B. F. Jacobs, B. Kyongo, M. Macharwas, and F. Muchemi. 2019. Primitive Old World monkey from the earliest Miocene of Kenya and the evolution of cercopithecoid biloph-odonty. Proceedings of the National Academy of Sciences 116:6051–6056.

Rasmussen, D. T., and M. Gutierrez. 2009. A mammalian fauna from the late Oligo-cene of northwestern Kenya. Palaeontographica Abteilung A 288:1–52.

Rasmussen, D. T., and E. L. Simons. 1991. The oldest Egyptian hyracoids (Mammalia: Pliohyracidae): New species of *Saghatherium* and *Thyrohyrax* from the Fayum. Neues Jahrbuch Für Geologie Und Paläontologie. Abhandlungen 182:187–209.

Rasmussen, D. T., and E. L. Simons. 1992. Paleobiology of the oligopithecines, the earliest known anthropoid primates. International Journal of Primatology 13:477–508.

Rathbun, G. B. 2009. Why is there discordant diversity in sengi (Mammalia: Afrotheria: Macroscelidea) taxonomy and ecology? African Journal of Ecology 47:1–13.

Ravel, A., M. Adaci, M. Bensalah, A.-L. Charruault, E. M. Essid, H. K. AMmar, W. Marzougui, M. Mahboubi, F. Mebrouk, G. Merzeraud, M. Vianey-Liaud, R. Tabuce, and L. Marivaux. 2016. Origine et radiation initiale des chauves-souris modernes: nouvelles découvertes dans l'Éocène d'Afrique du Nord. Geodiversitas 38:355–434.

Ravel, A., L. Marivaux, R. Tabuce, M. Adaci, M. Mahboubi, F. Mebrouk, and M. Bensalah. 2011. The oldest African bat from the early Eocene of El Kohol (Algeria). Naturwissenschaften 98:397–405.

Reyment, R. A. 1980. Biogeography of the Saharan Gretaceous and Paleocene epicontinental transgressions. Cretaceous Research 1:299–327.

Reyment, R. A., and R. V. Dingle. 1987. Palaeogeography of Africa during the Cretaceous Period. Palaeogeography, Palaeoclimatology, Palaeoecology 59:93–116.

Rögl, F. 1998. Palaeogeographic considerations for Mediterranean and Paratethys seaways (Oligocene to Miocene). Annalen Des Naturhistorischen Museums in Wien. Serie A Für Mineralogie Und Petrographie, Geologie Und Paläontologie, Anthropologie Und Prähistorie 99:279–310.

Russell, D. A., and M. A. Paesler. 2003. Environments of mid-Cretaceous Saharan dinosaurs. Cretaceous Research 24:569–588.

Sallam, H. M., and E. R. Seiffert. 2016. New phiomorph rodents from the latest Eocene of Egypt, and the impact of Bayesian "clock"-based phylogenetic methods on estimates of basal hystricognath relationships and biochronology. PeerJ 4:e1717.

Sallam, H. M., and E. R. Seiffert. 2020. Revision of Oligocene "*Paraphiomys*" and an origin for crown Thryonomyoidea (Rodentia: Hystricognathi: Phiomorpha) near the Oligocene–Miocene boundary in Africa. Zoological Journal of the Linnean Society 190:352–371.

Sallam, H. M., A. H. Sileem, E. R. Miller, and G. F. Gunnell. 2016. Deciduous dentition and dental eruption sequence of *Bothriogenys fraasi* (Anthracotheriidae, Artiodactyla) from the Fayum Depression, Egypt. Palaeontologia Electronica 19:1–17.

Samonds, K. E., I. S. Zalmout, M. T. Irwin, D. W. Krause, R. R. Rogers, and L. L. Raharivony. 2009. *Eotheroides lambondrano*, new middle Eocene seacow (Mammalia, Sirenia) from the Mahajanga Basin, northwestern Madagascar. Journal of Vertebrate Paleontology 29:1233–1243.

Sanders, W. J., J. Kappelman, and D. T. Rasmussen. 2004. New large-bodied mammals from the late Oligocene site of Chilga, Ethiopia. Acta Palaeontologica Polonica 49:365–392.

Schlesinger, G. 1913. Studien über die stammesgeschichte der Proboscidier. Jahrbuch Der Kaiserlich-Koniglichen Geologischen Reichsanstalt 62:87–182.

Schlosser, M. 1910. Über einige fossile Säugetiere aus dem Oligocän von Ägypten. Zoll. Anz. 35:500–508.

Schlosser, M. 1911. Beiträge zur Kenntnis der Oligozânen Land-Saugetiere aus dem Fayum: Ägypten. Beiträge Zur Paläontologie Und Geologie Österreich-Ungarns 24:51–167.

Scotese, C. R. 2021. An atlas of phanerozoic paleogeographic maps: The seas come in and the seas go out. Annual Review of Earth and Planetary Sciences 49:679–728.

Scotese, C. R., H. Song, B. J. W. Mills, and D. G. van der Meer. 2021. Phanerozoic paleotemperatures: The earth's changing climate during the last 540 million years. Earth-Science Reviews 215:103503.

Seiffert, E. R. 2007. A new estimate of afrotherian phylogeny based on simultaneous analysis of genomic, morphological, and fossil evidence. BMC Evolutionary Biology 7:224.

Seiffert, E. R. 2010. The oldest and youngest records of afrosoricid placentals from the Fayum Depression of northern Egypt. Acta Palaeontologica Polonica 55:599–616.

Seiffert, E. R. 2012. Early primate evolution in Afro-Arabia. Evolutionary Anthropology: Issues, News, and Reviews 21:239–253.

Seiffert, E. R., D. M. Boyer, J. G. Fleagle, G. F. Gunnell, C. P. Heesy, J. M. G. Perry, and H. M. Sallam. 2018. New adapiform primate fossils from the late Eocene of Egypt. Historical Biology 30:204–226.

Seiffert, E. R., S. Nasir, A. Al-Harthy, J. R. Groenke, B. P. Kraatz, N. J. Stevens, and A. R. Al-Sayigh. 2012. Diversity in the later Paleogene proboscidean radiation: A small barytheriid from the Oligocene of Dhofar Governorate, Sultanate of Oman. Naturwissenschaften 99:133–141.

Seiffert, E. R., and E. L. Simons. 2013. Last of the oligopithecids? A dwarf species from the youngest primate-bearing level of the Jebel Qatrani Formation, northern Egypt. Journal of Human Evolution 64:211–215.

Seiffert, E. R., E. L. Simons, D. M. Boyer, J. M. G. Perry, T. M. Ryan, and H. M. Sallam. 2010. A fossil primate of uncertain affinities from the earliest late Eocene of Egypt. Proceedings of the National Academy of Sciences 107:9712–9717.

Seiffert, E. R., E. L. Simons, W. C. Clyde, J. B. Rossie, Y. Attia, T. M. Bown, P. Chatrath, and M. E. Mathison. 2005. Basal anthropoids from Egypt and the antiquity of Africa's higher primate radiation. Science 310:300–304.

Seiffert, E. R., E. L. Simons, T. M. Ryan, T. M. Bown, and Y. Attia. 2007. New remains of Eocene and Oligocene Afrosoricida (Afrotheria) from Egypt, with implications for the origin(s) of afrosoricid zalambdodonty. Journal of Vertebrate Paleontology 27:963–972.

Seiffert, E. R., M. F. Tejedor, J. G. Fleagle, N. M. Novo, F. M. Cornejo, M. Bond, D. de Vries, and K. E. Campbell. 2020. A parapithecid stem anthropoid of African origin in the Paleogene of South America. Science 368:194–197.

Sen, S. 2013. Dispersal of African mammals in Eurasia during the Cenozoic: Ways and whys. Geobios 46:159–172.

Senut, B., and M. Pickford. 2021. Micro-cursorial mammals from the late Eocene tufas at Eocliff, Namibia. Communications of the Geological Survey of Namibia 23:90–160.

Shoshani, J., R. C. Walter, M. Abraha, S. Berhe, P. Tassy, W. J. Sanders, G. H. Marchant, Y. Libsekal, T. Ghirmai, and D. Zinner. 2006. A proboscidean from the late Oligocene of Eritrea, a "missing link" between early Elephantiformes and Elephantimorpha, and biogeographic implications. Proceedings of the National Academy of Sciences 103:17296–17301.

Sigé, B., J.-J. Jaeger, J. Sudre, and M. Vianey-Liaud. 1990. *Altiatlasius koulchii* n. gen., et sp., Primate omomyidé du Paléocène supérieur du Maroc, et les origines des

Euprimates. Palaeontographica. Abteilung A, Paläozoologie, Stratigraphie 214:31–56.

Sigogneau-Russell, D., M. Monbaron, and D. Russell. 1988. Découverte de mammifères dans le Mésozoïque moyen d'Afrique. Comptes Rendus de l'Académie Des Sciences. Série 2, Mécanique, Physique, Chimie, Sciences de l'univers, Sciences de La Terre 307:1045–1050.

Simons, E. L. 1962. Two new primate species from the African Oligocene. Postilla 64:1–12.

Simons, E. L. 1964. Notes. Society of Vertebrate Paleontology, News Bulletin 70:14–15.

Simons, E. L. 1965. New fossil apes from Egypt and the initial differentiation of Hominoidea. Nature 205:135–139.

Simons, E. L. 1968. Early Cenozoic mammalian faunas Fayum province, Egypt. Part I. African Oligocene mammals: Introduction history of study and faunal succession. Peabody Museum of Natural History, Yale University, 28:1–21.

Simons, E. L. 1989. Description of two genera and species of late Eocene Anthropoidea from Egypt. Proceedings of the National Academy of Sciences 86:9956–9960.

Simons, E. L., and T. M. Bown. 1985. *Afrotarsius chatrathi*, first tarsiiform primate (? Tarsiidae) from Africa. Nature 313:475–477.

Simons, E. L., and T. M. Bown. 1995. Ptolemaiida, a new order of Mammalia—with description of the cranium of *Ptolemaia grangeri*. Proceedings of the National Academy of Sciences 92:3269–3273.

Simons, E. L., S. Cornero, and T. Bown. 1998. The taphonomy of fossil vertebrate quarry L-41, upper Eocene, Fayum Province, Egypt. Geological Survey of Egypt Special Publication 75:785–791.

Simons, E. L., and D. T. Rasmussen. 1994. A remarkable cranium of *Plesiopithecus teras* (Primates, Prosimii) from the Eocene of Egypt. Proceeding of the National Academy of Sciences of the USA 91:9946–9950.

Simons, E. L., and D. T. Rasmussen. 1996. Skull of *Catopithecus browni*, an early tertiary catarrhine. American Journal of Physical Anthropology 100:261–292.

Simons, E. L., D. T. Rasmussen, and P. D. Gingerich. 1995. New cercamoniine adapid from Fayum, Egypt. Journal of Human Evolution 6:577–589.

Simpson, G. G. 1980. Splendid isolation: The curious history of South American mammals. Yale University Press, New Haven.

Solé, F., E. Essid, W. Marzougui, R. Temani, H. Khayati Ammar, M. Mahboubi, L. Marivaux, M. Vianey-Liaud, and R. Tabuce. 2016. New fossils of Hyaenodonta (Mammalia) from the Eocene localities of Chambi (Tunisia) and Bir el Ater (Algeria), and the evolution of the earliest African hyaenodonts. Palaeontologia Electronica 19:1–23.

Solé, F., J. Falconnet, and D. Vidalenc. 2015. New fossil Hyaenodonta (Mammalia, Placentalia) from the Ypresian and Lutetian of France and the evolution of the Proviverrinae in southern Europe. Palaeontology 58:1049–1072.

Solé, F., J. Lhuillier, M. Adaci, M. Bensalah, M. Mahboubi, and R. Tabuce. 2014. The hyaenodontidans from the Gour Lazib area (?early Eocene, Algeria): Implications concerning the systematics and the origin of the Hyainailourinae and Teratodontinae. Journal of Systematic Palaeontology 12:303–322.

Solé, F., M. Morlo, T. Schaal, and T. Lehmann. 2021. New hyaenodonts (Mammalia) from the late Ypresian locality of Prémontré (France) support a radiation of the hyaenodonts in Europe already at the end of the early Eocene. Geobios 66–67:119–141.

Springer, M. 2004. Molecules consolidate the placental mammal tree. Trends in Ecology & Evolution 19:430–438.

Springer, M. S., R. W. Meredith, J. E. Janecka, and W. J. Murphy. 2011. The historical biogeography of Mammalia. Philosophical Transactions of the Royal Society B: Biological Sciences 366:2478–2502.

Stevens, N. J., P. M. O'Connor, C. Mtelela, and E. M. Roberts. 2022. Macroscelideans (Myohyracinae and Rhynchocyoninae) from the late Oligocene Nsungwe formation of the Rukwa Rift Basin, southwestern Tanzania. Historical Biology 34:604–610.

Stevens, N. J., E. R. Seiffert, P. M. O'Connor, E. M. Roberts, M. D. Schmitz, C. Krause, E. Gorscak, S. Ngasala, T. L. Hieronymus, and J. Temu. 2013. Palaeontological evidence for an Oligocene divergence between Old World monkeys and apes. Nature 497:611–614.

Stromer, E. 1903. *Zeuglodon*-Reste aus dem oberen Mitteleocan des Fajum. Beiträge Zur Paläontologie Und Geologie Österreich-Ungarns Und Des Orients 15:65–100.

Sudre, J. 1979. Nouveaux Mammifères éocènes du Sahara occidental. Palaeovertebrata 9:83–115.

Tabuce, R. 2010. Solving the mystery of the enigmatic mammal *Helioseus insolitus*: A highly derived hyrax from the Eocene of Gour Lazib, Algeria. Journal of Vertebrate Paleontology 30:1A–198A.

Tabuce, R. 2018. New remains of *Chambius kasserinensis* from the Eocene of Tunisia and evaluation of proposed affinities for Macroscelidea (Mammalia, Afrotheria). Historical Biology 30:251–266.

Tabuce, R., A.-L. Charruault, M. Adaci, M. Bensalah, B. H. Ali, E. M. Essid, L. Marivaux, and M. Vianey-Liaud. 2011. The early Eocene radiation of Hyracoidea (Mammalia, Afrotheria): New fieldwork evidence from northwestern Africa. The World at the Time of Messel 161–162.

Tabuce, R., J.-J. Jaeger, L. Marivaux, M. Salem, A. A. Bilal, M. Benammi, Y. Chaimanee, P. Coster, B. Marandat, X. Valentin, and M. Brunet. 2012. New stem elephant-shrews (Mammalia, Macroscelidea) from the Eocene of Dur At-Talah, Libya: Palaeontology 55:945–955.

Tabuce, R., M. Mahboubi, and J. Sudre. 2001. Reassessment of the Algerian Eocene hyracoid *Microhyrax*. Consequences on the early diversity and basal phylogeny of the order Hyracoidea (Mammalia). Eclogae Geologicae Helvetiae 94:537–545.

Tabuce, R., L. Marivaux, M. Adaci, M. Bensalah, J.-L. Hartenberger, M. Mahboubi, F. Mebrouk, P. Tafforeau, J.-J. Jaeger. 2007. Early Tertiary mammals from North Africa reinforce the molecular Afrotheria clade. Proceedings of the Royal Society B 274:1159–1166.

Tabuce, R., L. Marivaux, R. Lebrun, M. Adaci, M. Bensalah, P.-H. Fabre, E. Fara, H. Gomes Rodrigues, L. Hautier, J.-J. Jaeger, V. Lazzari, F. Mebrouk, S. Peigné, J. Sudre, P. Tafforeau, X. Valentin, and M. Mahboubi. 2009. Anthropoid versus strepsirrhine status of the African Eocene primates *Algeripithecus* and *Azibius*: Craniodental evidence. Proceedings of the Royal Society B: Biological Sciences 276:4087–4094.

Tabuce, R., R. Sarr, S. Adnet, R. Lebrun, F. Lihoreau, J. E. Martin, B. Sambou, M. Thiam, and L. Hautier. 2020. Filling a gap in the proboscidean fossil record: A new genus from the Lutetian of Senegal. Journal of Paleontology 94:580–588.

Tassy, P. 1981. Le crâne de *Moeritherium* (Proboscidea, Mammalia) de l'Eocène de Dor El Talha (Libye) et le problème de la classification phylogénétique du genre dans les Tethytheria McKenna, 1975. Bulletin Du Muséum National d'Histoire Naturelle, Paris, 4 Sér., Sect. C 3:87–147.

Tassy, P. 2015. La phylogénie des proboscidiens (Mammalia); une question de méthode. Biosystema 30:81–97.

Thomas, H., J. Roger, S. Sen, and Z. Al-Sulaimani. 1988. Découverte des plus anciens "anthropoïdes" du continent arabo-africain et d'un primate tarsiiforme dans l'Oligocène du sultanat d'Oman. Comptes Rendus de l'Académie Des Sciences. Série 2, Mécanique, Physique, Chimie, Sciences de l'univers, Sciences de La Terre 306:823–829.

Thomas, H., J. Roger, S. Sen, C. Bourdillon-De-Grissac, and Z. Al-Sulaimani. 1989. Découverte de Vertébrés fossiles dans l'Oligocène inférieur du Dhofar (Sultanat d'Oman). Geobios 22:101–120.

Thomas, H., J. Roger, S. Sen, J. Dejax, M. Schuler, Z. A. Sulaimani, C. B. de Grissac, G. Breton, F. D. Broin, G. Camoin, H. Carpetta, R. P. Carriol, C. Cavelier, C. Chaix, J. Y. Crochet, G. Farjanel, M. Gayet, E. Gheerbrant, A. Lauriat-Rage, and D. Noel. 1991. Essai de reconstitution des milieux de sedimentation et de vie des primates anthropoides de l'Oligocene de Taqah (Dhofar, Sultanat d'Oman). Bulletin de La Société Géologique de France 162:713–724.

Thomas, H., S. Sen, J. Roger, and Z. Al-Sulaimani. 1991. The discovery of *Moeripithecus markgrafi* Schlosser (Propliopithecidae, Anthropoidea, Primates), in the Ashawq Formation (early Oligocene of Dhofar Province, Sultanate of Oman). Journal of Human Evolution 20:33–49.

Van Couvering, J. A., and E. Delson. 2020. African land mammal ages. Journal of Vertebrate Paleontology 40:e1803340.

Van der Made, J. 1999. Intercontinental relationship Europe-Africa and the Indian subcontinent; pp. 457–472 in G. E. Rossner and K. Heissig (eds.), The Miocene Land Mammals of Europe. Verlag Dr. Friedrich Pfeil, Munchen.

Vianey-Liaud, M., J.-J. Jaeger, J.-L. Hartenberger, and M. Mahboubi. 1994. Les rongeurs de l'Eocene d'Afrique nord-occidentale (Glib Zegdou (Algérie) et Chambi (Tunisie) et l'origine des Anomaluridae. Palaeovertebrata 23:93–118.

Voss, M., M. S. M. Antar, I. S. Zalmout, and P. D. Gingerich. 2019. Stomach contents of the archaeocete *Basilosaurus isis*: Apex predator in oceans of the late Eocene. PLOS ONE 14:e0209021.

Vrielynck, B., J. Dercourt, and N. Cottereau. 1994. Des seuils lithosphériques dans la Téthys. Comptes Rendus de l'Académie Des Sciences. Série 2. Sciences de La Terre et Des Planètes 318:1677–1685.

Wible, J. R., G. W. Rougier, M. J. Novacek, and R. J. Asher. 2007. Cretaceous eutherians and Laurasian origin for placental mammals near the K/T boundary. Nature 447:1003–1006.

Wood, A. E. 1968. Early Cenozoic mammalian faunas, Fayum Province, Egypt. Part II: The African Oligocene Rodentia. Peabody Museum of Natural History, Yale University, 28:23–105.

Yans, J., M. Amaghzaz, B. Bouya, H. Cappetta, P. Iacumin, L. Kocsis, M. Mouflih, O. Selloum, S. Sen, J.-Y. Storme, and E. Gheerbrant. 2014. First carbon isotope chemostratigraphy of the Ouled Abdoun phosphate Basin, Morocco; implications for dating and evolution of earliest African placental mammals. Gondwana Research 25:257–269.

Zalmout, I. S., and P. D. Gingerich. 2012. Late Eocene sea cows (Mammalia, Sirenia) from Wadi al Hitan in the western desert of Fayum, Egypt. Museum of Paleontology, University of Michigan, 37:1–158.

Zalmout, I. S., W. J. Sanders, L. M. MacLatchy, G. F. Gunnell, Y. A. Al-Mufarreh, M. A. Ali, A.-A. H. Nasser, A. M. Al-Masari, S. A. Al-Sobhi, A. O. Nadhra, A. H. Matari, J. A. Wilson, and P. D. Gingerich. 2010. New Oligocene primate from Saudi Arabia and the divergence of apes and Old World monkeys. Nature 466:360–364.

Zouhri, S., P. Gingerich, S. Adnet, E. Bourdon, S. Jouve, B. Khalloufi, A. Amane, N. Elboudali, J.-C. Rage, F. De Lapparent De Broin, A. Kaoukaya, and S. Sebti. 2018. Middle Eocene vertebrates from the sabkha of Gueran, Atlantic coastal basin, Saharan Morocco, and their peri-African correlations. Comptes Rendus Geoscience 350:310–318.

Zouhri, S., P. D. Gingerich, N. Elboudali, S. Sebti, A. Noubhani, M. Rahali, and S. Meslouh. 2014. New marine mammal faunas (Cetacea and Sirenia) and sea level change in the Samlat Formation, Upper Eocene, near Ad-Dakhla in southwestern Morocco. Comptes Rendus Palevol 13:599–610.

Zouhri, S., P. Gingerich, B. Khalloufi, E. Bourdon, S. Adnet, S. Jouve, N. Elboudali, A. Amane, J.-C. Rage, R. Tabuce, and F. de Lapparent de Broin. 2021. Middle Eocene vertebrate fauna from the Aridal Formation, Sabkha of Gueran, southwestern Morocco. Geodiversitas 43:121–150.

Zouhri, S., I. S. Zalmout, and P. D. Gingerich. 2022. New protosirenid (Mammalia, Sirenia) in the late Eocene sea cow assemblage of southwestern Morocco. Journal of African Earth Sciences 104516.

Taxa Index

Subject Index